Inter-Asterisk Exchange (IAX)

WILEY SERIES IN COMMUNICATIONS NETWORKING
& DISTRIBUTED SYSTEMS

Series Editors: David Hutchison, *Lancaster University, Lancaster, UK*
Serge Fdida, *Université Pierre et Marie Curie, Paris, France*
Joe Sventek, *University of Glasgow, Glasgow, UK*

The 'Wiley Series in Communications Networking & Distributed Systems' is a series of expert-level, technically detailed books covering cutting-edge research, and brand new developments as well as tutorial-style treatments in networking, middleware and software technologies for communications and distributed systems. The books will provide timely and reliable information about the state-of-the-art to researchers, advanced students and development engineers in the Telecommunications and the Computing sectors.

Other titles in the series:

Wright: *Voice over Packet Networks* 0-471-49516-6 (February 2001)
Jepsen: *Java for Telecommunications* 0-471-49826-2 (July 2001)
Sutton: *Secure Communications* 0-471-49904-8 (December 2001)
Stajano: *Security for Ubiquitous Computing* 0-470-84493-0 (February 2002)
Martin-Flatin: *Web-Based Management of IP Networks and Systems* 0-471-48702-3 (September 2002)
Berman, Fox, Hey: *Grid Computing. Making the Global Infrastructure a Reality* 0-470-85319-0 (March 2003)
Turner, Magill, Marples: *Service Provision. Technologies for Next Generation Communications* 0-470-85066-3 (April 2004)
Welzl: *Network Congestion Control: Managing Internet Traffic* 0-470-02528-X (July 2005)
Raz, Juhola, Serrat-Fernandez, Galis: *Fast and Efficient Context-Aware Services* 0-470-01668-X (April 2006)
Heckmann: *The Competitive Internet Service Provider* 0-470-01293-5 (April 2006)
Dressler: *Self-Organization in Sensor and Actor Networks* 0-470-02820-3 (November 2007)
Berndt: *Towards 4G Technologies: Services with Initiative* 0-470-01031-2 (March 2008)
Jacquenet, Bourdon, Boucadair: *Service Automation and Dynamic Provisioning Techniques in IP/ MPLS Environments* 0-470-01829-1 (March 2008)
Minei/Lucek: *MPLS-Enabled Applications: Emerging Developments and New Technologies, Second Edition*
0-470-98644-1 (April 2008)
Gurtov: *Host Identity Protocol (HIP): Towards the Secure Mobile Internet* 0-470-99790-7 (June 2008)

Inter-Asterisk Exchange (IAX) Deployment Scenarios in SIP-Enabled Networks

Mohamed Boucadair
France Telecom, France

A John Wiley and Sons, Ltd, Publication

This edition first published 2009
© 2009 John Wiley & Sons, Ltd

Registered office
John Wiley & Sons Ltd, The Atrium, Southern Gate, Chichester, West Sussex, PO19 8SQ, United Kingdom

For details of our global editorial offices, for customer services and for information about how to apply for
permission to reuse the copyright material in this book please see our website at www.wiley.com.

Library of Congress Cataloging-in-Publication Data

Boucadair, Mohamed.
   Inter-asterisk exchange (IAX) : deployment scenarios in SIP-enabled networks / Mohamed
Boucadair.
       p. cm.
   Includes bibliographical references and index.
   ISBN 978-0-470-77072-6 (cloth)
1. Internet telephony. 2.   Computer network protocols. I. Title.
   TK5105.8865.B68 2009
   004.69′5–dc22   2008037697

A catalogue record for this book is available from the British Library.

ISBN 9780470770726 (H/B)

Typeset in 10/12pt Times by Thomson Digital, Noida, India.
Printed in Great Britain by CPI Antony Rowe, Chippenham, Wiltshire.

Contents

Part One: IAX Protocol Specifications

Part Three: Deployment Scenarios in SIP-Based Environments

Foreword

Voice over IP service offerings have been deployed for quite some years, hence strengthening the Internet as the privileged support for conveying a wide variety of value-added multimedia streams. But the design and the operation of such multimedia services at the scale of the Internet remains complex and demands a high level of expertise.

Indeed, the deployment of multimedia service offerings demands high levels of quality, reliability and security that can only be reached by means of a set of advanced capabilities that need to be supported in different regions of the network.

For example, the establishment of a multimedia session requires the ability to (1) reliably identify the customer who generated the request to access the multimedia contents, (2) make sure this customer is entitled to access the contents he has requested, (3) (assuming access to contents has been granted to the customer) make sure the multimedia contents are delivered to the customer with the relevant level of quality (possibly yielding the design and the enforcement of a Quality of Service policy, which will rely upon the combination of traffic identification and classification capabilities, traffic conditioning and scheduling capabilities, etc.) and security (preserving the identity and the integrity of the customer, preserving the confidentiality of the transactional exchanges related to payment of the contents, preserving the delivery of multimedia contents from any kind of (distributed) denial of service attack, etc.).

The amount of signalling information that needs to be exchanged between the relevant devices (hosts, routers, platforms) before a multimedia session can be established has become increasingly significant as access to (rich) multimedia content and services has become more common.

This factual situation could become even worse in forthcoming Next Generation Network (NGN) environments based upon the so-called IP Multimedia Subsystem (IMS) service command platform, where interactions between transport-control functions (e.g. the Resource Allocation Control Function), service-control functions (e.g. the maintenance of user profiles) and application-control functions (e.g. maintenance of the contents) are required, let alone the signalling traffic that needs to be exchanged between the end-user functions and each of these control functions before the communication can be established.

Within such architectures, access to contents may very well take several tens of seconds (if not minutes), and this is clearly incompatible with customers' expectations, which are generally reflected through the notion of Quality of Experience.

Several protocols have been designed by the Internet Engineering Task Force to signal and control multimedia sessions, including the Session Initiation Protocol (SIP) and the Media Gateway Control Protocol (MGCP).

These protocols are often described as a generic, flexible toolbox, from which different features will be selected and activated depending on the nature of the service (e.g. private extensions to the SIP protocol in case media authorisation capabilities are required) and/or the networking context (e.g. extensions to the SIP protocol for symmetric response routing), yielding further design complexity, if not extra overhead.

In addition, these signalling protocols need to be activated in conjunction with other protocols that will convey the multimedia data once the session is established. For example, the Session Description Protocol (SDP) is often solicited by SIP to describe the contents of the data that will be conveyed by yet another protocol such as the Real-Time Transport Protocol (RTP).

Furthermore, multimedia service designs that make use of private addressing assume address translation mechanisms which are hardly compatible with these signalling protocols, raising additional security issues.

This is where the Inter-Asterisk eXchange (IAX) protocol jumps in, as a proposal to address the current issues that have been briefly mentioned above.

The IAX protocol seems to be one of those good, promising ideas that come from the open-source community, which aims at dramatically facilitating the deployment and the operation of multimedia services, by combining control and media transport facilities in one protocol with great flexibility.

There is undoubtedly an ever growing interest from the (VoIP) service provider community for IAX (whose version 2 is currently being discussed within the IETF), and this book, which provides an in-depth description of the IAX protocol and usage, is the most comprehensive effort I'm aware of.

It indeed not only covers the IAX protocol details and machinery, but also, rightfully, its application in various contexts (including pan-provider and IPv6 environments).

This book has been written by a remarkably knowledgeable expert of the domain who has gained the recognition of the international research community for many years, thanks to his numerous and awarded publications in IP QoS and interdomain traffic engineering areas, among others.

Put simply, this book is a must-have for anyone who's involved in the development, the design and/or the operation of advanced, value-added multimedia service offerings.

Christian Jacquenet
Director of Strategic Programs for IP Networks and Services, France Telecom

Terminology and Definitions

The terminology used within this book is aligned with the one defined in [IAX] and [SIP]. To ease the reading of this book, hereafter are listed the most useful definitions as introduced in the aforementioned documents:

- *Address of Record (AoR)*: An address-of-record is an URI (Uniform Resource Identifier) that points to a domain with a location service that can map the URI to another URI where the user might be available. Typically, the location service is populated through registrations.
- *Back-to-Back User Agent (B2BUA)*: A back-to-back user agent is a logical entity that receives a request and processes it as a user agent server. In order to determine how the request should be answered, it acts as a user agent client and generates requests.
- *Call*: This term denotes some communication between peers, generally set up for the purposes of a multimedia conversation.
- *Client*: A client is any network element that sends IAX/SIP requests and receives IAX/SIP responses.
- *Home Domain*: The domain providing service to a SIP/IAX user. Typically, this is the domain present in the URI in the address-of-record of a registration.
- *Initiator, Calling Party, Caller*: The party initiating a session with a call initiation request. A caller retains this role from the time it sends the initial call set-up request that established a dialog until the termination of that dialog.
- *Invitee, Invited User, Called Party, Callee*: The party that receives call set-up request for the purpose of establishing a new session.
- *Request*: A SIP/IAX message sent from a client to a server, for the purpose of invoking a particular operation.
- *Response*: A SIP/IAX message sent from a server to a client, for indicating the status of a request sent from the client to the server.
- *Call Identifier*: A call leg is marked with two unique integers, one assigned by each peer involved in creating the call leg. It is used to uniquely identify the call.
- *Dialplan*: This term denotes a set of rules for associating provided names and numbers with a particular called party;
- *Frame*: The atomic communication unit between two IAX peers;
- *Information Element*: A discrete data unit appended to an IAX frame which specifies user or call-specific data;

- *Number*: The Calling and Called Numbers are a set of digits and letters identifying a call originator and the desired terminating resource.
- *Peer*: Within this book this term denotes a host, device or terminal that supports IAX protocol;
- *Registrant*: A registrant is a peer that makes registration requests in order to advertise the address of a resource.
- *Registrar*: A registrar is a server that processes registration requests and places the information it receives in those requests into the location service;
- *Username*: A username is a string used for identification purposes.
- *Location Server*: An entity which is able to store, maintain and retrieve the location of a given resource in the context of a given service.
- *IP Telephony Domain*: An administrative perimeter which delimits the boundaries of an IP telephony service managed by a given Service Provider.
- *Proxy Server*: An entity which is able to handle received request and proceeds to appropriate telephony routing actions.
- *Address of Contact (AoC)*: An AoC is identified by a FQDN (Fully Qualified Domain Name) or an IP address and a port number to be used to reach a given user.
- *End-point*: Denotes en extremity involved in a given communication
- Adjacent domains: Denotes two IP telephony domains which are neighbours. Interconnection agreements are settled so as to allow exchange of traffic between them.

[SIP]   Rosenberg, J., Schulzrinne, H., Camarillo, G., Johnston, A., Peterson, J., Sparks, R., Handley, M., and E. Schooler, "*SIP: Session Initiation Protocol*", RFC 3261, June 2002

[IAX]   Spencer, M., Shumard, K., Capouch, B., and E. Guy, "*IAX2: Inter-Asterisk eXchange Version 2*", draft-guy-iax, Work In Progress

Acronyms and Abbreviations

This book uses the following acronyms:

ADSI	Analog Display Service Interface
ADSICPE	Analog Display Services Interface CPE
AES	Advanced Encryption Standard
ANI	Automatic Number Identification
AUTII MDS	Authentication methods
B2BUA	Back-to-Back User Agent
BGP	Border Gateway Protocol
CALLED CTX IE	Called Context Information Element
CALLED NB IE	Called Number Information Element
CALLING NB IE	Calling Number Information Element
CALLING PRES IE	Calling Presentation Information Element
CALLING TON IE	Calling Type Of Number Information Element
CALLNO IE	Call Number Information Element
CLID	Calling Line Identification
CODEC	Compression/Decompression
CODEC PREFS IE	Compression Decompression Preference Information Element
CPE	Customer Premise Equipment
DDDS	Dynamic Delegation Discovery System
DNID	Dialed Number Identification
DNS	Domain Name Service
DNS RR	Domain Name Service Resource Record
DNS SRV	DNS Service Record
DUNDI	Distributed Universal Number Discovery
E2U	E.164 to URI
ENCKEY	Encryption Key
ENUM	Telephone Number Mapping
FC	Full Cone
FQDN	Fully Qualified Domain Name
FW BLOCL DESC	Firmware Block Description
FWBLOCK DATA	Firmware block Data
GROW WG	Global Routing Operations Working Group

HNT	Hosted NAT Traversal
IAX	Inter-Asterisk Exchange
IAX2	IAX version 2
ICE	Interactive Connectivity Establishment
IE	Information Element
IESG	Internet Engineering Steering Group
IETF	Internet Engineering Task Force
IGP	Interior Gateway Protocol
IM	Instant Messaging
IMS	IP Multimedia Subsystems
IP	Internet Protocol
IPSec	IP Security
IPv6	IP version 6
Iseqno	Incoming Sequence Number
IS-IS	Intermediate System-Intermediate System
ITAD	IP Telephony Administrative Domain
MD5	Message Digest 5
MGW	Media Gateway
NAPTR	Naming Authority Pointer
NAT	Network Address Translation
NAT-PT	Network Address Translation - Protocol Translation
OOO	Out of Order
Oseqno	Outgoing Sequence Number
OSPF	OSPF: Open Shortest Path First
P2P	Peer to Peer
P2PSIP	Peer to Peer Session Initiation Protocol
PABX	Private Access Branch Exchange
PAT	Port Address Translation
PBX	Private Branch Exchange
PKTS	Packets
PLMN	Public Land Mobile Network
POTS	Plain Old Telephone Service
PSTN	Public Switched Telephone Network
QoS	Quality of Service
RELOAD	REsource LOcation And Discovery
RR	Resource Record
RR OOO	Frames Received Out Of Order
RSA	Rivest, Shamir and Adleman
RTCP	Real-time Transport Control Protocol
RTP	Real-Time Transport Protocol
SBC	Session Border Controller
SDP	Session Description Protocol
SGW	Signalling Gateway
SIP	Session Initiation Protocol
SIPPING WG	Session Initiation Proposal Investigation Working Gorup
SRTP	Secure RTP

STUN	Simple Traversal of UDP Through NATs
TCP	Transport Control Protocol
THIG	Topology Hiding Internetworking Gateway
TISPAN	Telecoms & Internet converged Services & Protocols for Advanced Networks
TLS	Transport Layer Security
ToIP	Telephony over IP
TRIP	Telephony Routing Over IP
TURN	Traversal Using Relay NAT
UA	User Agent
UAC	User Agent Client
UAS	User Agent Server
UDP	User Datagram Protocol
URI	Uniform Resource Identifier
VoIP	Voice over IP
WG	Working Group

Acknowledgement

To Saâdia, with all my Love.

To my parents, my brothers and my sisters.
To my friends…

Many thanks to Pierrick Morand and Yoann Noisette for the fruitful discussions we had earlier
to the writing of this book ....

1

Introduction

1.1 General Introduction

Voice over IP (VoIP) is a privileged field of service innovation within an effervescent telecommunication environment. Most service providers (SPs) have started to migrate or at least plan on migrating their PSTN (Public Switched Telephone Network) infrastructure to an IP-based one. Within this context, IMS (IP Multimedia Subsystems, [IMS]) and TISPAN (Telecoms & Internet Converged Services & Protocols for Advanced Networks, [TISPAN]) architectures have been specified and promoted by the 3GPP (3rd Generation Partnership Project) community to meet service providers' requirements, in particular to ease fixed–mobile convergence, and to accelerate the PSTN renewal and replacement of TDM (Time Division Multiplexing) by IP.

IMS and TISPAN architectures use SIP (Session Initiation Protocol, [SIP]) as the VoIP signalling protocol. This choice was motivated by the popularity of the protocol and its emergence within the IETF (Internet Engineering Task Force) community. SIP was specified, by the IETF community and then adopted by 3GPP, as a protocol which is suitable for controlling heterogeneous multimedia sessions over IP.

In earlier stages of telephony over IP (ToIP) deployments and in a context where H.323 [H.323] had started to attract service providers, SIP was rapidly adopted by them owing to its richness, its flexibility and its claimed simplicity compared to H.323. This adoption was motivated by the dynamic created within IETF around SIP and its associated extensions. Indeed, SIP has been promoted as a simple and extensible protocol. This openness of the protocol has been 'exploited' by protocol designers, who advocate for introducing SIP to solve any kind of problem (e.g. establishment of IPSec (IP Security, [IPSEC]) tunnels). Note that the aforementioned SIP simplicity is no longer a valid argument today. For instance, SIP documentation is more than 1200 pages (additional interesting statistics may be found at rfc3261.net). This makes it difficult to implement interoperable equipment and systems. The complexity is also related to the base SIP specification itself, which include 628 occurrences of 'MUST', 342 of 'SHOULD' and 377 of 'MAY' occurrences. The specifications are therefore ambiguous and detailed design of algorithms and protocol behaviours is left to the implementers. This leads to the emergence of various implementations which are not interoperable.

Inter-Asterisk Exchange (IAX): Deployment Scenarios in SIP-Enabled Networks   Mohamed Boucadair
© 2009 John Wiley & Sons, Ltd

In addition to the above-mentioned complexity, SIP suffers from several other hurdles, such as the difficulties of crossing NAT (Network Address Translation, [NAT]) and firewall boxes, the operational difficulty of setting up media sessions (due to dynamic RTP (Real-Time Transport Protocol, [RTP]) port numbers assignment policy), complications arising from its path-decoupled nature (since service providers needs to insert an intermediate node in both the signalling and the media path, for instance for access-control purposes), the emergence of SIP-unfriendly boxes (which are not standardised and break the SIP end-to-end paradigm), and the need to deploy a SIP Protocol Suite (SDP (Session Description Protocol, [SDP]), RTP, RTCP (Real-Time Transport Control Protocol, [RTP]), STUN (Simple Traversal of UDP Through NATs, [STUN]), TURN (Traversal Using Relay NAT, [TURN]), ICE (Interactive Connectivity Establishment, [ICE]), etc.) almost as large as the famous 'H.323 umbrella'![1]

Service providers should take into account these drawbacks in order to investigate how the SIP protocol, companion protocols and associated architectures may be enhanced (which is not an easy task, because some of the SIP complications are caused by its design choices, such as the presence of IP-related information in the SIP/SDP bodies, which is from an architectural viewpoint a bad practice), or whether there are viable alternatives which meet service providers' requirements and do not suffer from these critical 'SIP pains'.

From this perspective, this book presents the IAX (Inter-Asterisk Exchange, [IAX]) protocol as a possible candidate to solve SIP complications. Introduction scenarios and methods for-easing the introduction of IAX into SIP-based networks are elaborated, and a clear strategy to 'exploit' the advantages of both IAX and SIP for the delivery of multimedia services, especially conversational ones, is described.

1.2 On Voice over IP and Telephony over IP

Within this book, VoIP and ToIP are used interchangeably. The subtle differences between these two services are ignored, since our area of investigation is orthogonal to legal constraints (such as legal intercept and emergency calls) and service-packaging issues. Furthermore, this book does not assume any specific conversational services, even if a focus is put on audio and video ones. Indeed, the discussions and analyses conducted here should apply to whatever type of session IAX and/or SIP is used to manage.

1.3 Context

This section sketches the context within which Telcos are evolving. This context should be carefully considered and taken into account when proposing solutions to service providers' requirements.

1.3.1 Proliferation of Middleboxes

Middleboxes, particularly NAT boxes, have been ignored for a long period by the IETF and no standardisation effort has been undertaken within that organisation. The motivation for this

---

[1] ITU has specified a standard for audio, video and data communications over IP, called the H.323 recommendation. This recommendation is commonly referred to as an umbrella recommendation, since it includes parts of Q.931, RAS, T120, H.245, RTP, RTCP, G.723, G.711, G.728, H.261, H.263, etc.

position is to avoid the specification of systems which are against the 'end-to-end principle'. This desertion of NATs by the IETF has led to the emergence of heterogeneous implementations of the NAT function, which is perceived by network architects as a nightmare. Later, the IETF edited several documents to analyse available implementations and to identify the problems caused by the presence of such a function in the network. Recently, a working group has been chartered to investigate and to specify the required behaviours of NAT. In the meantime, service providers have started to integrate this function in their bundled CPE (Customer Premise Equipment) in order to easily extend the scope of their service offerings to various pieces of equipment present in the home network. As a result, those service providers have been confronted with NAT traversal issues for some of their service offerings, especially telephony over IP. To solve this issue, additional modules are embedded in the home gateways and new service nodes are introduced in the IP Telephony Administrative Domain. This additional complexity may be avoided if the protocols used have been designed to easily cross NATs.

The IETF has failed to 'shape' NAT function and to promote interoperable and open implementations. Besides this failure, the IETF has promoted protocols which suffer from rudimentary design flaws such as interference between OSI layers (e.g. SIP, which carries IP-related information). This interference, especially in the context of SIP, is a big problem when considering deployment scenarios. Indeed, several protocols, procedures and functions have been introduced to ease SIP NAT traversal.

1.3.2 IP Exhaustion Problem

The service provider community is aware of the exhaustion of public IPv4 addresses. In this context, the community was mobilised in the past to adopt a 'promising' solution, in particular with the definition of IPv6 (Internet Protocol Version 6). Nevertheless, this solution is not globally activated by service providers, for financial and strategic reasons. In the meantime, these service providers are not indifferent to the alarms recently emitted by the IETF. G. Huston introduced and promoted an extrapolation model to forecast the exhaustion date of IPv4 addresses managed by IANA (Internet Assigned Numbers Authority). This effort indicates that if the current tendency of consumption continues as it is, the date of the exhaustion of IPv4 addresses of IANA's pool would be 2011, while that of the RIRs (Regional Internet Registry) would be 2012. In order to solve this exhaustion problem, service providers should investigate and activate short-term solutions and continue to offer their IP-based service offerings. One of the most investigated solutions is denoted 'Provider NAT' (also called 'Double NAT'). This solution proposes to introduce an additional level of NAT, hosted at service provider perimeter.

In order to deliver SIP-based calls in the presence of Provider NAT boxes, service providers should be aware of the underlying IP infrastructure so as to implement appropriate ALGs (Application-Level Gateway). At least the modification of SIP messages should be enforced: first at the Home NAT and then at Provider NAT. If no such ALG is enabled, no communication may be established. This constraint is 'heavy', since it assumes a vertical integration (that is, no functional separation between the service provider and the underlying IP network provider) and that the same administrative entity administers both service and network infrastructure.

The next challenge is to avoid deploying 'heavy' architectures to solve the IP exhaustion problem.

1.3.3 Migration to IPv6

The IETF has been working for several years on migration issues related to IPv6. Several service providers envisage adopting IPv6 as the new connectivity protocol for many reasons, such as the abundance of addresses, to take advantage of the routing hierarchy or to benefit from the native auto-configuration features supported by IPv6. Furthermore, IPv6 has been adopted as the main IP protocol in the context of several architectures, such as IMS. This mid–long-term objective should be taken into account when designing new architectures to be deployed by service providers for the delivery of their service offerings. As far as conversational services are concerned, the problem is not related to the delivery of the service over IPv6, but is to ensure interworking between IPv4 and IPv6 realms. From a SIP perspective, new adaptation functions should be activated so as to ease the establishment of successful sessions between heterogeneous user agents (that is, IPv4 and IPv6).

1.3.4 Lightweightness and Optimisation of CAPEX and OPEX

Session Border Controllers have been designed and promoted by several vendors in order to meet a set of technical and legal requirements expressed by service providers. These SIP-unfriendly nodes are not standardised and are proprietary. Several interoperability and service support issues have been identified by service providers during their validation phase; the introduction of these nodes into operational networks should also be assessed and evaluated from a CAPEX (Capital Expenditure) and an OPEX (Operational Expenditure) perspective. Furthermore, the presence of SBC nodes in the service delivery chain introduces additional technical problems and constraints on QoS (Quality of Service) and robustness. Several functions supported by these SBCs are due to SIP design choices. A lightweight SBC implementation would be envisaged, so as to optimise CAPEX and OPEX. This requirement is not only valid for the service access segment but also for the overall service architecture.

1.3.5 Avoid the Overspecification Phenomenon

A balanced approach should be adopted when specifying a given protocol. Openness of the protocol is good practice, but this should not increase the complexity of implementation tasks and induce interoperability issues. Furthermore, clear requirements and objectives should drive the design of a given protocol. SIP is an example of a protocol which suffers from the 'overspecification phenomenon'. Concretely, several features of the protocol are not required for the delivery of telephony services. A more pragmatic approach would be privileged. As an example of this phenomenon, designers encounter problems deciding on which criteria the authentication procedure should be enforced. Several options and alternatives have been investigated, and new SIP headers have even been introduced. Another example is the ambiguity of the routing process. To clarify this issue, a new RFC has been edited by the IETF.

Besides this specification ambiguity, the protocol is not optimised for telephony services. The overall architecture (mainly SIP, SDP and RTP) is not designed to ease correlation between signalling data and media streams. These two stacks are managed separately. As a consequence, additional nodes are required to maintain additional states and to implement this correlation between signalling messages and media flows. Bandwidth optimisation concerns are also valid since RTP encloses an overhead which is not required in the context of telephony services. As an example, this book describes IAX as a means to ease correlation between signalling message and media streams, and also to optimise required bandwidth for exchanging media streams.

1.3.6 Interconnection Issues

In order to extend the scope of a given telephony service beyond the administrative boundaries of a single domain, service providers should cooperate and interconnect. This interconnection will encourage the enforcement of global reachability and allow local customers to place their calls to destinations attached to remote service providers' domains. The underlying complexity required to offer this global reachability should be hidden, and handled between service providers. Furthermore, to implement this service, routing policies should be enforced so as to avoid PSTN realms and reduce interconnection fees. For these reasons, appropriate methods should be investigated and activated, such as telephony routing protocols. Moreover, appropriate signalling protocols should be activated to place interdomain calls and avoid exposing sensitive data (e.g. internal service topology) to external parties.

1.4 Enhancement Strategies to Solve SIP Issues

It is commonly agreed that SIP encounters a plethora of technical hurdles. These hurdles are mainly caused by its design choices. Indeed, SIP does not 'follow' the OSI layers and uses information which belongs to underlying layers. For these reasons, the SIP community within the IETF has been obliged to investigate new solutions to solve these technical problems. Starting from a simple and attractive base, SIP has become a complex and heavy protocol to implement. SIP should not be reduced to these technical problems but should be seen from a wider perspective. It offers interesting features such as routing, forking and so on. These features are not supported by IAX, for instance, and are part of the service providers' requirements.

Various enhancement methodologies may be adopted to solve SIP complications. Besides the patch-based approach adopted by the IETF, this book proposes a novel solution which takes advantage of SIP features in appropriate service segments and activates an alternative protocol for the delivery of conversational services where SIP is not considered a lightweight answer. This approach avoids introducing into operational networks architectures and protocols which are not considered lightweight from a manageability perspective.

1.5 IAX: Towards Lightweight Telephony Architectures

IAX stands as an interesting alternative besides classical protocols, deployed nowadays by service providers for their conversational service offerings (e.g. H.323 and SIP). This book illustrates how IAX could fulfil a large set of service providers' requirements and even bring more to their architectures, mainly the native support of traditional services. IAX is a path-coupled protocol that is used for both signalling and media-control operations. Moreover, it provides interesting features such as management of signalling and media transfer, support for native provisioning functions and firmware maintenance. IAX is a simple protocol, which has the advantage of being IP version agnostic, leading to avoidance of NAT traversal complications. This issue represents a real asset, as NAT boxes are nowadays a tremendous challenge in conversational architectures and services and require additional patches, especially in home gateway equipment and the first service equipment (notably 'Hosted NAT Traversal' facility). Moreover, this combined simplicity and completeness makes it germane to avoid resorting to a SIP Protocol Suite (SIP, SDP, RTP, RTCP, STUN, ICE, TURN...).

The IAX protocol offers significant features unavailable in other existent VoIP signalling protocols. Apart from its simplicity, the main characteristics of the IAX protocol are listed below:

- IAX is transported over UDP (User Datagram Protocol) using a single port number. The default IAX port is 4569.
- The IAX registration philosophy is the same as the SIP one. An IAX registrant should contact a registrar server with specific messages. Contact information is then retrieved by the registrar server and stored in its system within a time period.
- IAX couples signalling and media paths. The decoupling is possible once the connection has been successfully established. This characteristic is denoted 'path-coupled' protocol, in contrast with the 'path-decoupled' approach assumed by SIP.
- IAX does not require a new protocol for the exchange of media streams. It handles media streams itself. Various media types may be sent by IAX: voice, video, image, text, HTML and so on.
- IAX defines reliable and unreliable messages. IAX-unreliable messages are media flows which are not acknowledged nor retransmitted if lost in the network. IAX reliability is ensured for control messages thanks to several IAX application identifiers maintained by the involved parties. Reliable messages should be acknowledged; if not, these messages are retransmitted.
- NAT traversal is not a nightmare anymore with IAX. No IP addresses are enclosed in IAX signalling messages.
- IAX defines a set of messages used to monitor the status of the network. These messages can be exchanged during or outside an active call.
- IAX offers the means to check whether a remote call participant is alive or not.
- Native IP security methods can be deployed jointly with IAX. IAX allows exchange of shared keys. It may be used either with plain text or in conjunction with encryption mechanisms like AES (Advanced Encryption Standard, [AES]). Unlike SIP, no confusion is raised by identity-related information used to enforce authentication.
- IAX authentication is implemented thanks to the exchange of authentication requests, which enclose a security challenge. This authentication challenge should be answered by the remote peer and encrypted according to the adopted encryption method. If encryption negotiation has failed, the call should be terminated.
- IAX provides a dedicated scheme to provision IAX devices through a specific procedure and IAX messages.
- IAX allows a procedure to check the availability of a new firmware version for a given device type. The encoding of firmware binary blocks is specific to IAX devices and is out of the scope of the IAX communication protocol itself.
- IAX can be easily deployed to provide heterogeneous calls between IPv4 and IPv6 realms.
- And so on.

The activation of IAX in an operational network will simplify current architectures and therefore there will be no need to introduce expensive and SIP-unfriendly nodes. The proposed IAX introduction scenario is accompanied by an extension to SDP to allow smooth migration and media optimisation.

1.6 IAX and Standardisation

IAX was developed in the context of the Asterisk Project. In earlier stages of that project, no documentation was edited, according to the principle of 'documentation is the code'. But recently an individual Internet draft was submitted to the IETF. It was sent to RFC Editor so as to be adopted as an individual submission according to the IETF RFC publication process. After a first evaluation phase, this publication request was forwarded to the IESG (Internet Engineering Steering Group). This board has made a decision and sees no problem in publishing 'IAX: Inter-Asterisk eXchange Version 2' (draft-guy-iax-04.txt) as an IETF Informational RFC. Furthermore, IESG thinks that this work is related to IETF work done in SIP, MMUSIC (Multiparty Multimedia Session Control) and AVT (Audio/Video Transport) working groups, but this does not prevent publishing.

The IAX Internet draft is currently in the RFC Editor queue. Once editing checking has been undertaken by the RFC Editor, this Internet draft will be published as an Information RFC. This track should not be confused with the 'standard track'. The advantage of being published within the IETF is being able to disseminate the protocol and allow a wide publication of the document among the Internet community and then among service providers and Telcos. Moreover, the publication of the IAX Internet draft as an RFC is understood, as IAX is not against activities conducted within IETF working groups.

Additional information related to this Internet draft may be found at datatracker.ietf.org/idtracker/draft-guy-iax.

1.7 Rationale

To allow the introduction of IAX, the adopted methodology in this book is incremental: first to analytically show the added value of the IAX protocol compared to existing ones, and then to propose viable deployment scenarios to assess the behaviour of the protocol in operational networks. Indeed, IAX can be seen as a complement, for instance at the access segment of service providers' conversational services, or even a replacement at mid-term of the existing protocols in their conversational service platforms and architectures. IAX could help in getting rid of problems linked to NAT owing to its native support: no more heavy ALGs or HNT (Hosted NAT Traversal) mechanisms. This would decrease, if not suppress, the need for expensive SBCs, which moreover wouldn't need to perform TH (Topology Hiding) operations anymore.

In particular, this book aims to introduce IAX as a viable alternative which can solve operational issues related to the deployment of conversational services. This book does not aim to provide detailed specifications regarding how to enable IAX at the access segment, nor to exhaustively identify required functions, but only to sketch viable scenarios by which we can benefit from IAX capabilities within the operational environment. This book takes the position that IAX should not be seen as replacement for SIP in all use cases, but that it should be introduced in situations where it is better than SIP.

1.8 What This Book is Not

This book does not provide an overview of SIP. An abundance of papers, books and position papers has already been produced regarding SIP. Readers are invited to refer to this literature if required. This book does refer to SIP specifications and practices when necessary.

This book does not put IAX against SIP, but presents an alternative where IAX and SIP are deployed together to meet a service provider's requirements and ease delivery of their service offerings. The focus is on the service provider itself and not on the underlying technological means used to deploy a given service. IAX and SIP are presented as a toolbox. The use of this toolbox is left to the service providers themselves. Lightweightness and ease of manageability to handle networking issues should be privileged.

1.9 Structure of the Book

This book is structured into three major parts as described below.

1.9.1 Part One: IAX Protocol Specification

Part One describes the IAX Uniform Resource Identifier (URI) scheme and provides examples of URIs. ENUM (E.164 Telephone Number Mapping) architectures and the use of IAX in ENUM-enabled realms are also provided. Then IAX protocol objects ('full', 'mini' and 'meta' frames) are introduced. IAX information elements and IAX requests and their function objectives are also presented. Several taxonomy methods have been detailed. This first part then focuses on IAX connectivity considerations, especially the used transport protocol, call multiplexing, IAX reliability and IAX timers. Finally, Part One provides a set of examples of supported IAX operations such as registration, call management, call setup, call monitoring and so on.

1.9.2 Part Two: Discussion and Analysis

Part Two focuses on various uses of the IAX protocol and its capability to offer advanced services, to handle some painful networking issues and to be easily extended so as to cover a large set of conversational features.

Chapter 9 focuses first on the ability of the IAX protocol to implement a CODEC negotiation between remote IAX peers and the support of the 'on-fly' CODEC negotiation feature. It describes in particular the ability of IAX to manage video sessions. A section is dedicated to an enhancement to the IAX protocol which optimises the number of exchanged control messages between two IAX peers. Furthermore, the ability of the IAX protocol to support presence services and instant messaging is analysed. Overviews are given of IAX and its native support of the topology hiding function, and of the support of IAX issues when mobile IP is deployed. Finally, this chapter highlights how some miscellaneous features, such as call transfer, call forward, fax and so on are supported when IAX is deployed.

Chapter 10 is dedicated to IAX deployment in a multiserver environment. It focuses first on the means to enforce discovery of IAX resources. Two categories of these means are identified and then described: static and dynamic. An overview is then provided of end-to-end call setup in the presence of several IAX servers in the path. Load balancing features in an IAX environment are discussed, and implementation options described. Additionally, the need for service providers to enforce both path-coupled and path-decoupled architectures is given. Then the path-coupled characteristic of IAX and its ability to be enhanced to support a path-decoupled mode are highlighted. Finally, this chapter provides a brief overview of the inability of current IAX specifications to achieve 'forking' features which avoid telephony routing loops. Route symmetry issues and the need for the signalling response path to follow the same route as the request path are also mentioned.

Chapter 11 discusses NAT traversal issues when the IAX protocol is activated for the delivery of conversational services. It presents the IP exhaustion problem and two solutions to it. IAX can be activated in the context of these solutions, and does not pose additional technical problems. Unlike SIP, IAX is powerful for NAT traversal and the delivery of reliable communications.

Chapter 12 focuses on P2P (peer-to-peer) service offerings and the applicability of IAX to delivering P2P conversational services. A new architecture based on native IP capabilities is introduced. New IAX objects and messages are defined to support distributed conversational services. The proposed architecture is multicast-based distributed architecture and does not require deployment of heavy DHT (Distributed Hash Table) infrastructure, nor centralized nodes. It is suitable for implementation for corporate customers since it offers flexibility and simplifies required configuration operations.

Chapter 13 discusses the impact of the introduction of IPv6 on IAX-based service offerings. Several scenarios are evaluated and discussed. This chapter shows that the activation of IAX in an IPv6-enabled environment will not encounter major problems.

Finally, Chapter 14 presents the notion of the 'IP telephony administrative domain' and gives a macroscopic functional view of a telephony service platform. Furthermore, it identifies two deployment scenarios for SBC nodes: access and interconnection deployment. An overview of the motivations for introducing SBC nodes into SIP architectures is provided, and two categories of motivation are identified and described: technical problems and legal require-ments. A functional decomposition of an SBC node and both media and signalling considera-tions are given in this chapter. Additionally, it lists several functions supported by SBC nodes and gives a brief overview of each one. Finally, it checks the applicability of SIP-oriented SBC functions in IAX-based service architectures.

1.9.3 Part Three: Deployment Scenarios in SIP-Based Environments

Part Three is dedicated to elaborate candidate scenarios for introducing IAX into an SIP-based environment.

Chapter 15 argues for the need to enhance current service architectures and to simplify these architectures to avoid complications related to SIP. These complications are induced by SIP design choices and additional protocols must be activated to solve them. The activation of these protocols introduces new manageability issues that should be taken into account by service providers when specifying their architectures. This chapter also presents the adopted method-ology to enhance the current SIP-based architectures and lists a set of facts to be taken into account. These items should drive the specification effort of an enhancement solution. Moreover, a set of requirements to be considered when proposing new solutions is described and a brief comparison between IAX and SIP is also included. Finally, a set of scenarios for activating IAX in operational networks are identified.

Chapter 16 provides numerous call flows to illustrate the behaviour of the proposed IAX–SIP interworking function. This chapter shows that the introduction of such a function into operational networks should ease the traversal of middleboxes. It also introduces an extension to SDP to allow end-to-end bandwidth optimisation.

Finally, Chapter 17 describes a validation scenario to assess the feasibility of the proposed strategy for introduction of IAX into an SIP-enabled environment. This validation scenario does not aim to assess the performance of the proposed solution but only to provide a 'proof of concept' system. Required configuration operations are provided in this chapter, together with excerpts from configuration files.

References

[AES] US Department of Commerce/NIST, 'FIPS-197, Announcing the Advanced Encryption Standard', November 2001.

[H.323] ITU-T Recommendation H.323, 'Packet-based Multimedia Communications Systems', International Telecommunication Union (ITU-T), November 2000.

[IAX] Spencer, M., Shumard, K., Capouch, B. and Guy, E., 'IAX2: Inter-Asterisk eXchange Version 2', draft-guy-iax-04, work in progress.

[ICE] Rosenberg, J., 'Interactive Connectivity Establishment (ICE): A Methodology for Network Address Translator (NAT) Traversal for Offer/Answer Protocols', draft-ietf-mmusic-ice-12, October 2006.

[IMS] Camarillo, G. and Garcia-Martin, M.A., *The 3G IP Multimedia Subsystem – Merging the Internet and the Cellular Worlds*, John Wiley and Sons, Ltd., 2005.

[IPSEC] Kent, S. and Atkinson, R., 'Security Architecture for the Internet Protocol', RFC 2401, November 1998.

[NAT] Holdrege, M. and Srisuresh, M.,'Protocol Complications with the IP Network Address Translator', RFC 3027, January 2001.

[RTP] Schulzrinne, H., Casner, S., Frederick, R. and Jacobson, V.,'RTP: A Transport Protocol for Real-Time Applications', RFC 1889 (proposed standard), January 1996.

[SBC] Hautakorpi, J. et al., 'Requirements from SIP (Session Initiation Protocol) Session Border Control Deployments', draft-camarillo-sipping-sbc-funcs-05.

[SDP] Handley, M., Jacobson, V. and Perkins, C.,'SDP: Session Description Protocol', RFC 5466, July 2006.

[SIP] Rosenberg, J., Schulzrinne, H., Camarillo, G., Johnston, A., Peterson, J., Sparks, R. et al., 'SIP: Session Initiation Protocol', RFC 3261, June 2002.

[STUN] Rosenberg, J., Weinberger, J., Huitema, C. and Mahy, R.,'STUN – Simple Traversal of User Datagram Protocol (UDP) Through Network Address Translators (NATs)', RFC 3489, March 2003.

[TISPAN] TISPAN, 'Telecommunications and Internet Converged Services and Protocols for Advanced Networking, NGN Release 1', TR180001, 2006.

[TURN] Rosenberg, J. et al., 'Traversal Using Relay NAT (TURN)', work in progress.

Further Reading

Poikselka, M. and Mayer, G., *'The IMS: IP Multimedia Concepts and Services'* 3rd Edition, John Wiley and Sons, Ltd., November 2008.

Sinnreich, H. and Johnston, A., *Internet Communications Using SIP: Delivering VoIP and Multimedia Services with Session Initiation Protocol*, 2nd Edition, John Wiley and Sons, Ltd., August 2006.

VoIP RFC Watch, http://rfc3261.net/.

2

The IAX Protocol at a Glance

This section lists a set of valid questions that may be asked by a reader who is not familiar with the IAX protocol. Answers are concise and condensed since the ambition is not to provide full clarifications but only to furnish sampler elements, which the author hopes will encourage the reader to enjoy the remaining parts of this book.

Note that all these questions are discussed and answers elaborated in further chapters (especially in the first part of this book), but not always in the same order as they appear in this chapter.

2.1 What Does IAX Stand For?

IAX is the abbreviation of **I**nter-**A**sterisk e**X**change [AST].

2.2 Is IAX Specific to the Asterisk Platform?

No. IAX is not specific to the Asterisk platform. IAX can be used independently of Asterisk to control the signalling and the media between two endpoints (also denoted as peers or user agents (UAs)).

Asterisk is a multichannel PBX (Private Branch Exchange) which supports several VoIP (voice over IP) signalling protocols, such as SIP (Session Initiation Protocol, [SIP]), H.323 [H.323] and IAX.

2.3 What is the Difference between IAX2 and IAX?

In VoIP literature, IAX2 is used to denote the IAX protocol version 2. Within this book, we use IAX to denote all versions of the IAX protocol, including IAX2.

The current IAX version 2 specifications require that the same version is supported by all call participants.

2.4 Why another New VoIP Protocol?

The existent VoIP protocols suffer from a plethora of technical problems, such as NAT (Network Address Translator) traversal. New companion protocols are introduced within the

Inter-Asterisk Exchange (IAX): Deployment Scenarios in SIP-Enabled Networks   Mohamed Boucadair
© 2009 John Wiley & Sons, Ltd

IETF (Internet Engineering Task Force) to solve some of these problems. Examples of these companion protocols are ICE (Interactive Connectivity Establishment, [ICE]), STUN (Simple Traversal of UDP through NATs, [STUN]) and TURN (Traversal Using Relay NAT, [TURN]).

The introduction of these protocols means additional complexity for service designers and implementers. Furthermore, interoperability and interworking issues become critical for service providers and solution integrators.

IAX contributes to solving several technical issues encountered in current VoIP signalling protocols. Additional features are supported by IAX, unlike existent VoIP protocols.

2.5 How Does IAX Solve VoIP Pains?

Unlike other VoIP signalling protocols, IAX uses a single port number to send and/or receive both media and control data.

Unlike SIP, IAX does not require the inclusion of IP-related information in its call setup requests.

Unlike SIP with RTP (Real-Time Transport Protocol, [RTP]), IAX uses a static port number to send and/or receive media flows. The same session is used for both signalling and media traffic exchange. A multiplexing capability is supported by IAX to distinguish ongoing sessions.

2.6 How is Calls Multiplexing Achieved?

In order to multiplex the active calls managed by a local peer, IAX introduces at the application layer a specific identifier, called **Source Call Number**, to unambiguously identify local sessions. This identifier is not a port number or an IP address which belongs to underlying layers (as with OSI layer), but is managed at the application level.

2.7 And What About Demultiplexing?

For the received flows, a given IAX peer uses an identifier called **Destination Call Number** at the application layer.

2.8 What Port Number Does IAX Use?

For all IAX sessions, a single port number is used. IANA has assigned a port number dedicated to IAX; this number is **4569**. It is used for both signalling and media traffic exchange.

2.9 What Transport Protocol Does IAX Use?

All IAX messages are sent over UDP (User Datagram Protocol, [UDP]). Reliability features are supported at the application level.

2.10 Is IAX a Reliable Protocol?

Yes. IAX ensures the reliability of some of its messages at the application layer. In particular, IAX ensures that control messages are acknowledged.

2.11 How Does IAX Ensure Reliability?

IAX defines reliable and unreliable messages.

Unreliable messages are mainly media flows which are not acknowledged nor retransmitted if lost when sent over the network. IAX reliability is ensured for control messages, owing to several IAX application identifiers maintained by the peers. Reliable messages should be acknowledged; if not, these messages are retransmitted. The order of received messages is achieved by exploiting a timestamp enclosed in them.

2.12 Is there an IAX Registration Procedure?

IAX supports the registration procedure, even if it is considered an optional feature. The IAX registration philosophy is the same as for SIP. An IAX registrant contacts a registrar server with specific messages. The contact information is then retrieved by the registrar server and stored in its system within a time period.

2.13 Does IAX Registration Differ from SIP Registration?

SIP uses a dedicated message called **REGISTER** to notify a registrar server about its AoC (Address of Contact). An AoC may be an FQDN (Fully Qualified Domain Name) or an IP address and a port number. If a registered UA is behind an NAT, this AoC must be modified by an ALG (Application-Level Gateway) so as to replace its information with some which is more pertinent.

Unlike with SIP, IAX registration messages do not enclose any IP-related information (i.e. IP address and a port number). The registrar server extracts the source IP-related information and stores it. This information will be used to contact the registrant user agent. Consequently, no ALG is required to modify IAX messages.

2.14 How Are Media Streams Transported in IAX?

Unlike SIP, IAX does not require a new protocol to exchange media streams. Numerous media types may be sent by IAX: voice, video, image, text, HTML (Hypertext Markup Language) and so on. IAX uses an optimised protocol header (**4 bytes**) to send audio messages.

Note that media streams use the same port number as for control/signalling messages.

2.15 Is CODEC Negotiation Supported by IAX?

Yes. When issuing the first call-establishment request, a given IAX peer can indicate one or several CODECs (Compression/Decompression) it supports. The remote IAX peer must select one of these CODECs and indicates it in its response message.

2.16 Is On-Fly CODEC Change Possible During a Call?

Yes. IAX allows on-fly CODEC change. The new CODEC to be used is indicated in a specific frame.

2.17 IAX: a Path-Coupled or Decoupled Protocol?

IAX couples signalling and media paths. Therefore the media path is the same as the signalling one. Nevertheless, the decoupling may be enforced once the session has been successfully established. Indeed, intermediary IAX servers may be removed from the call legs and IAX traffic to be exchanged directly between remote endpoints.

This design choice is mainly motivated by the need to ease establishment of sessions and help endpoints to place their calls. Some NAT traversal problems may be encountered if those servers are not involved in the call leg.

2.18 Can IAX be Aware of the Status of the Network Load?

Yes. Unlike SIP and RTP, IAX defines a set of dedicated messages used to monitor the status of the network. These messages may be exchanged either during or outside an active IAX call.

2.19 What About Security?

IAX is a point-to-point communication protocol. Native IP security protocols and architectures can be deployed jointly with IAX.

Furthermore, IAX allows exchange of shared keys. IAX may be used either with plain text or in conjunction with encryption mechanisms such as AES (Advanced Encryption Standard, [AES]) which are based on a shared secret. The IAX authentication procedure is implemented by an exchange of authentication requests which enclose a security challenge. This authentication challenge should be answered by the remote peer and encrypted according to the adopted encryption method. If encryption negotiation fails, the call should be terminated.

2.20 Could IAX Devices Be Managed?

Yes. IAX provides a dedicated scheme to provision IAX devices through a specific procedure and associated IAX messages. The provisioning data template and contents are out of the scope of the IAX protocol itself.

2.21 Is Firmware Version Updating Supported by IAX?

Yes. IAX implements a dedicated procedure to check the availability of a new firmware version for a given device type. The encoding of firmware binary blocks is specific to IAX devices and is out of the scope of the IAX communication protocol itself.

2.22 Can IAX Be Extended to Offer New Features?

IAX supports a set of features. These features can be enriched through the introduction of new messages and new informational elements. Additional features may be deployed by using new information elements in existing IAX messages.

2.23 How is an IAX Resource Identified?

An IAX resource is identified by a URI (Uniform Resource Identifier, [URI]), which indicates the location of the IAX resource and any useful information such as the name and port number.

2.24 What Does an IAX URI Look Like?

Examples of IAX URIs include:

- **iax:iax.orange-ftgroup.com/PMO**
- **iax:pm2vpmo.orange-ftgroup.com:4569/PMO**
- **iax:pmo@pm2vpmo.orange-ftgroup.com:4569/PMO**
- **iax:pmo@[2001:688:::1]:4569/PMO_user.**

2.25 Is it Possible to Set a Call Involving Several IAX Servers?

IAX is a VoIP protocol used to set one or several call legs within the context of an end-to-end call. Some call legs can be established within another VoIP signalling protocol such as SIP or H.323. Furthermore, several IAX servers can then be involved in one or more call legs belonging to same end-to-end call.

2.26 Is it Possible to Discover the Location of an IAX Resource?

Numerous methods can be implemented to discover which IAX server is managing a particular IAX resource:

- Static configuration: a given IAX server is configured to route its call setup requests to an adjacent IAX server for a given telephone prefix.
- DUNDi protocol, [DUNDi].
- TRIP protocol, [TRIP].
- ENUM service, [ENUM].

2.27 What Is DUNDi?

DUNDi (Distributed Universal Number Discovery) is a peer-to-peer (P2P) telephony extension discovery, as defined in [DUNDI]. When activating DUNDi, several peers can exchange the number extension they support. Owing to this trust relationship, a peer within this community can retrieve the location of an IAX resource present in the network.

This protocol can be used with any VoIP signalling protocol.

2.28 What Is TRIP?

TRIP is a telephony routing protocol introduced by the IETF. This protocol aims at discovering and announcing telephony prefixes to TRIP peers. It is similar to the BGP (Border Gateway Protocol, [BGP]) for interdomain routing and IS-IS (Interior System–Interior System, [ISIS]) for intradomain routing.

This telephony routing protocol can be used with any VoIP signalling protocol (modulo some extensions).

2.29 What Is ENUM?

ENUM is a DNS-based approach to resolving the IP location of a given telephony resource. IAX service registration within ENUM system is defined in [IAXENUM].

References

[AES]   US Department of Commerce/NIST, 'FIPS-197, Announcing the Advanced Encryption Standard', November 2001.

[AST]   http://www.asterisk.org/.

[BGP]   Rekhter, Y., Li, T. and Hares, S., 'A Border Gateway Protocol 4 (BGP-4)', RFC 4271, January 2006.

[DUNDI]   Spencer, M., 'Distributed Universal Number Discovery (DUNDi)', draft-mspencer-dundi-01, October 2004.

[ENUM]   Faltstrom, P. and Mealling, M., 'The E.164 to Uniform Resource Identifiers (URI) Dynamic Delegation Discovery System (DDDS) Application (ENUM)', RFC 3761, April 2004.

[H323]   ITU-T Recommendation H.323, 'Packet-Based Multimedia Communications Systems', International Tele-communication Union (ITU-T), November 2000.

[IAX]   Spencer, M., Shumard, K., Capouch B. and Guy E., 'IAX2: Inter-Asterisk eXchange Version 2', draft-guy-iax-04, work in progress.

[IAXENUM]   Guy, E., 'IANA Registration for IAX Enumservice', draft-ietf-enum-iax-02, work in progress.

[ICE]   Rosenberg, J., 'Interactive Connectivity Establishment (ICE): A Methodology for Network Address Translator (NAT) Traversal for Offer/Answer Protocols', draft-ietf-mmusic-ice-12, October 2006.

[ISIS]   Oran, D., 'OSI IS-IS Intra-domain Routing Protocol', RFC 1142, February 1990.

[RTP]   Schulzrinne, H., Casner, S., Frederick, R. and Jacobson, V., 'RTP: A Transport Protocol for Real-Time Applications', RFC 1889 (proposed standard), January 1996.

[SIP]   Rosenberg, J., Schulzrinne, H., Camarillo, G., Johnston, A., Peterson, J. Sparks, R., et al., 'SIP: Session Initiation Protocol', RFC 3261, June 2002.

[STUN]   Rosenberg, J., Weinberger, J., Huitema, C. and Mahy R., 'STUN – Simple Traversal of User Datagram Protocol (UDP) through Network Address Translators (NATs)', RFC 3489, March 2003.

[TRIP]   Rosenberg, J. et al., 'Telephony Routing over IP (TRIP)', RFC 3219, January 2002.

[TURN]   Rosenberg, J. et al., 'Traversal Using Relay NAT (TURN)', work in progress.

[UDP]   Postel, J., 'User Datagram Protocol', RFC 768, August 1980.

[URI]   Berners-Lee, T., Fielding, R. and Masinter L., 'Uniform Resource Identifier (URI): Generic Syntax', STD 66, RFC 3986, January 2005.

Further Reading

Blanchet, M., *Asterisk Essentials: A Practical Guide to Open Source Voice Over IP*, November 2008, ISBN 978-0-470-06914-1.

Camarillo, G. and García-Martín, M., *The 3G IP Multimedia Subsystem (IMS): Merging the Internet and the Cellular Worlds*, 3rd Edition, September 2008, ISBN 978-0-470-51662-1.

Poikselka, M. and Mayer, G., *The IMS: IP Multimedia Concepts and Services*, 3rd Edition, November 2008, ISBN: 978-0-470-72196-4.

Sinnreich, H. and Johnston, A.B. *Internet Communications Using SIP: Delivering VoIP and Multimedia Services with Session Initiation Protocol*, 2nd Edition, August 2006, ISBN: 978-0-471-77657-4.

Part One

IAX Protocol Specifications

Part One of this book describes IAX protocols: the structure of its frames, its objects, its messages and its functional behaviours. This part does not replace the IAX specification document. Protocol implementers are invited to refer to the IAX specifications for more details and compliance clauses.

Part One is structured as follows:

- *Chapter 3*: describes the IAX Uniform Resource Identifier (URI) scheme and provides some examples of IAX URIs. Also introduces ENUM (*E.164 Telephone Number Mapping*) architectures and the use of IAX in ENUM-enabled realms.
- *Chapter 4*: describes the IAX protocol objects, especially, 'full', 'mini' and 'meta' frames. Also presents IAX informational elements and examples of traces.
- *Chapter 5*: describes IAX requests and their function objectives. Several taxonomy methods are also detailed.
- *Chapter 6*: focuses on IAX connectivity considerations, especially the transport protocol, call multiplexing, IAX reliability and IAX timers.
- *Chapter 7*: describes a set of examples of supported IAX operations, such as registration, call management, call setup, call monitoring and so on.

3

IAX Uniform Resource Identifier

3.1 Introduction

Customers should be able to reach every telephone number independently of the location of the remote destination. By location, we mean the VoIP Service Provider (VSP) it is attached to. Concretely, customers should have access to a large set of destination telephony numbers, everywhere in the world and independently of the originating and terminating telephony service providers and whatever the scheme of the remote customer identifier format is (for example, E.164 numbers [E.164] such as **+33231759231** or **+1245698466**). In the context of VoIP deployment and PSTN (Public Switched Telephone Network) renewal, new signalling protocols have been promoted and disseminated (for example, SIP (Session Initiation Protocol, [SIP]), MGCP (Media Gateway Control Protocol, [MGCP]), H.323, [H323], etc.). These protocols do not use all E.164 to implement the numbering and therefore to identify subscribed customers to their offered services. New numbering and user-identification schemes have been specified. These schemes are specific to each signalling protocol (SIP, IAX, H.323, etc.). Nevertheless, a plethora of these protocols rely on the notion of a Uniform Resource Identifier (URI), which provides a means for identifying an abstract or physical resource [RFC3986] within the network. A URI must follow strict syntax and parsing rules. In addition to the rules defined in the basic URI specifications, each signalling protocol defines its recommendations regarding the structure of the specific URI to be used to invoke the service deployed based on that protocol.

In this chapter, we describe URIs specific to the IAX protocol. Syntax and validation rules are described in Section 3.2. Section 3.4 lists a set of criteria for comparing IAX URIs and assessing whether or not two IAX URIs point to the same IAX resource. Section 3.5 is dedicated to ENUM (E.164 Telephone Number Mapping, [ENUM]), a key architecture which eases the convergence of service invocation by promoting a user-identification process. Concretely, a single resource is provisioned in DNS (Domain Name System), but several services may be 'attached'/associated with that resource. Examples of these services are VoIP, SMTP (Simple Mail Transfer Protocol, [SMTP]), PSTN, HTTP (Hypertext Transfer Protocol) and so on. More details and information are provided in Section 3.5.6. In such a system, a telephone number is the single pointer to diverse services but may use alternative schemes to invoke them.

Inter-Asterisk Exchange (IAX): Deployment Scenarios in SIP-Enabled Networks   Mohamed Boucadair
© 2009 John Wiley & Sons, Ltd

3.2 Format of IAX Uniform Resource Identifiers

According to [RFC3986], a URI is *'an identifier consisting of a sequence of characters matching the syntax rule named <URI>. It enables uniform identification of resources via a separately defined extensible set of naming schemes. How that identification is accomplished, assigned, or enabled is delegated to each scheme specification.'*. Examples of valid URIs include:

- FTP (File Transfer Protocol) URI: **ftp://ftp.test.com/rfc/rfc3986.txt**
- HTTP URL: **http://www.ietf.org/rfc/rfc3986.txt**
- LDAP (Lightweight Directory Access Protocol) URI: **ldap://21.22.23.24:85/dc=com? objectClass?one**
- SMTP URI: **mailto:test@example.com**
- Newsgroup URI: **news:comp.test.www.servers.unix**
- PSTN number or SIP 'tel' URI: **tel: + 33231759231**
- Telnet: **telnet://21.22.23.24:80/**

As far as IAX is concerned, an IAX URI identifies a resource capable of communicating using IAX protocol and manipulates its associated objects. Such an identifier encloses sufficient information to initiate a call (and generally an IAX session) towards the aforementioned resource. An IAX resource can be associated with the IAX server through which the call should be routed.

The general IAX URI [IAX] format is characterised in Table 3.1.

3.3 Examples of IAX Uniform Resource Identifiers

In order to familiarise the reader with IAX URIs, a set of examples of valid IAX URIs is provided below.

In Table 3.2, the listed IAX URIs serve the same user, named **pmo**. This user is IAX-enabled and is reachable at **pm2vpmo.orange-ftgroup.com**. The corresponding IP address of **pmo**'s machine is IPv4-capable and can be reached (from an IP connectivity perspective) at

Table 3.1   IAX URI format

```
An IAX URI must follow this format:
    'iax2:'[username@]host[:port][/number|name[?context]]
Where:
• username: a chain of strings used for identification purposes. This portion
  of the URI is optional;
• host: contains the IP contact information for the IAX subscriber. It may be
  filed as an FQDN or as an IPv4 or an IPv6 address. If an IPv6 address is used, it
  must be enclosed within brackets. The host part of an IAX URI is mandatory;
• port: a numeric value to indicate the port number used for binding the IAX
  service on the host identified by the host part. This part of the IAX URI is
  optional. If not present, the default value is 4569;
• number | name: a telephone extension number or a name identifying the IAX
  resource at the indicated host. This part of the IAX URI is optional;
• context: characterises the context in which this resource is identified.
  Call routing logic may exploit this part of a URI. This part of the IAX URI is
  optional.
```

Table 3.2 Examples of IAX URIs

```
IAX-URI.1.   iax:pm2vpmo.orange-ftgroup.com/PMO
IAX-URI.2.   iax:iax.orange-ftgroup.com/PMO
IAX-URI.3.   iax:pm2vpmo.orange-ftgroup.com:4569/PMO
IAX-URI.4.   iax:21.56.36.36:4569/pMo?work
IAX-URI.5.   iax:21.56.36.36/PmO
IAX-URI.6.   iax:pmo@pm2vpmo.orange-ftgroup.com:4569/212144546?invite
```

21.56.36.36. The port number to which this IAX service is bound in **pmo**'s host is the default
IAX port number: **4569**.

The interpretation of the IAX URIs in Table 3.2 is as follows:

- IAX-URI.1: **PMO** is IAX-enabled and can be reached at the machine identified by the FQDN
  (Fully Qualified Domain Name) **pm2vpmo.orange-ftgroup.com**. The port number to be
  used to initiate an IAX session is the IAX default port number, **4569**.
- IAX-URI.2: **PMO** is IAX-enabled and can be reached by contacting the server identified by
  the FQDN **iax.orange-ftgroup.com**. The port number to be used to initiate an IAX session is
  the IAX default port number, **4569**.
- IAX-URI.3: **PMO** is IAX-enabled and can be reached at **pm2vpmo.orange-ftgroup.com**.
  The UDP port to be used in the context of an IAX session destined to this IAX resource is **4569**.
- IAX-URI.4: **pMo** is IAX-enabled and can be reached at **21.56.36.36**. The port number to be
  used is **4569**. Unlike in the previous example, a context is associated with this resource.
  Indeed, the context of this IAX resource is **work**. This means that invoking this resource may
  be conditioned by the value of the context in which an IAX session might occur.
- IAX-URI.5: **PmO** is IAX-enabled and can be reached using IAX at **21.56.36.36**. The port
  number to be used is the IAX default port number, **4569**.
- IAX-URI.6: **pmo** is IAX-enabled and is reached at **pm2vpmo.orange-ftgroup.com**. The UDP
  port number to be used is **4569**. The number extension within the invite context is **212144546**.

Following the definition provided in Table 3.1, an IAX URI may be reduced to its mandatory
parts; that is, to the **host** part, as illustrated in the examples in Table 3.3.

The interpretation of the IAX URIs in Table 3.3 is as follows:

- IAX-URI.7: this URI encloses only the mandatory part of an IAX URI. It refers to an
  IAX resource which is reachable at **pm2vpmo.orange-ftgroup.com**. The port number to
  be used to initiate an IAX call towards this IAX resource is the IAX default port
  number, **4569**.

Table 3.3 Examples of IAX URIs (including only mandatory parts)

```
IAX-URI.7.          iax:pm2vpmo.orange-ftgroup.com
IAX-URI.8.          iax:21.56.36.36:4569
IAX-URI.9.          iax:21.56.36.36
IAX-URI.10.         iax:21.56.36.36/583219972
IAX-URI.11.         iax:pm2vpmo.orange-ftgroup.com/455483268
IAX-URI.12.         iax:pmo@pm2vpmo.orange-ftgroup.com/4554832686141
```

- IAX-URI.8: this URI refers to an IAX-enabled resource which is reachable at **21.56.36.36**. The port number to be used to establish an IAX session is **4569**. No username or resource name is specified in this IAX URI.
- IAX-URI.9: this URI encloses only the mandatory part of an IAX URI. It refers to an IAX-enabled resource reachable at **21.56.36.36**. The port number to be used is the IAX default port number, **4569**. No username or resource name is specified in this IAX URI.
- IAX-URI.10: this IAX URI refers to an IAX resource associated with the number extension **583219972**. This resource is IAX-enabled and is reachable at **21.56.36.36**. The port number to be used to initiate an IAX session is the IAX default port number, **4569**.
- IAX-URI.11: this IAX URI refers to an IAX resource associated with number extension **455483268**. This resource is IAX-enabled and is reachable at **pm2vpmo.orange-ftgroup. com**. The port number to be used is the IAX default port number, **4569**.
- IAX-URI.12: this IAX URI refers to an IAX resource identified by a user named **pmo**. This resource is IAX-enabled. It is reached at **pm2vpmo.orange-ftgroup.com**. The port number to be used to initial a call toward this resource is the IAX default port number, **4569**. The number extension is **4554832686141**.

If **PMO**'s machine has an available IPv6 connectivity, the following URIs are also valid IAX ones (Table 3.4). Note that in these examples, the IPv6 addresses **2001:688::1** and **::ffff:21.22.23.24** are enclosed between brackets so as to avoid confusion with ':', used to separate an IPv4 address and a port number. These brackets are not part of the IPv6 address itself. If a port number is part of the IAX URI enclosing an IPv6 address, then this port number must appear outside the delimiting ']'.

Table 3.4 Example of IAX URIs embedding IPv6 addresses

IAX-URI.13.	iax:[2001:688::1]:4569/PMO?work
IAX-URI.14.	iax:[::ffff:21.22.23.24]:7854/PMO

The first example illustrates an IPv6-embedded IAX URI. This IPv6 address is identified by the delimiting '[' and ']' symbols. The interpretation of the IAX URI is similar to an IPv4-embeded IAX URI. The second example involves a specific IPv6 address, **::ffff:21.22.23.24**, which is called an 'IPv4-mapped IPv6 address'. This type of IPv6 address has the first 80 bits set to **0** and the last 32 bits representing an IPv4 address (appearing in a dotted decimal format). The remaining 16 bits are used to distinguish IPv4-compatible IPv6 addresses from IPv4-mapped IPv6 addresses. A value of **ffff** is assigned to IPv4-mapped IPv6 addresses, and **0000** to IPv4-compatible IPv6 addresses. The latter are deprecated by the standards, since current IPv6 transition mechanisms do not use these addresses anymore (for more details, refer to [RFC4291]).

3.4 Comparing IAX Uniform Resource Identifiers

IAX URIs are used by 'location servers', registrars or other servers as the main 'key' to route IAX calls towards the IP contact information associated with a given IAX URI. The IP contact information denotes an IP address together with a port number. If no port number is specified in the IAX URI, a default value (**4569**) is used to initiate or to accept an IAX call. In order to achieve routing operations, the aforementioned entities should achieve some comparison operations so as to avoid confusion and not route an ongoing call to an incorrect endpoint. In order to meet this requirement, IAX defines a set of rules to determine whether two given IAX URIs are equivalent or not (that is, identify the same IAX resource). These rules are listed below:

- *Rule 1: Usernames must be identical.* Usernames are case sensitive. The IAX URIs in Table 3.5 are not equivalent because **PMO** and **pMo** are not identical.

Table 3.5   Examples of nonequivalent IAX URIs (case sensitive)

```
IAX-URI.1.    iax:PMO@pm2vpmo.orange-ftgroup.com
IAX-URI.2.    iax:pMo@pm2vpmo.orange-ftgroup.com
```

The IAX URIs in Table 3.6 are also distinct since the username parts are different.

Table 3.6   Examples of nonequivalent IAX URIs (distinct usernames)

```
IAX-URI.3.  iax:Levis_the_healthy@pm2vpmo.orange-ftgroup.com
IAX-URI.4.  iax:tagante@pm2vpmo.orange-ftgroup.com
```

- *Rule 2: The port numbers must be the same.* If omitted, the port number is considered to be equal to the default value, **4569**. The IAX URIs in Table 3.7 are not equivalent because the port number associated with the first URI is **6556** and the one associated with the second URI is **4512**.

Table 3.7   Examples of nonequivalent IAX URIs (distinct port numbers)

```
IAX-URI.1.    iax:PMO@pm2vpmo.orange-ftgroup.com:6556
IAX-URI.2.    iax:PMO@pm2vpmo.orange-ftgroup.com:4512
```

But the two IAX URIs in Table 3.8 are equivalent, because the first one does not enclose a port number and so the default value is assumed, while the second one use the default number, **4569**.

Table 3.8   Examples of nonequivalent IAX URIs (same port numbers)

```
IAX-URI.3.    iax:PMO@pm2vpmo.orange-ftgroup.com
IAX-URI.4.    iax:PMO@pm2vpmo.orange-ftgroup.com:4569
```

- *Rule 3: Two **host** parts are considered to be equal if their DNS resolution results in the same IP address.* A **host** encoded as an FQDN is considered to be equivalent to an IP address if and only if the latter is a result of a DNS query (i.e. an **A** RR (Resource Record) in the context of IPv4, or an **AAAA** RR when IPv6 is enabled, because IPv6 addresses are represented in the

Table 3.9   Examples of equivalent IAX URIs

```
IAX-URI.1.      iax:pm2vpmo.orange-ftgroup.com/PMO
IAX-URI.2.      iax:pm2vpmo.orange-ftgroup.com/pmo
IAX-URI.3.      iax:pm2vpmo.orange-ftgroup.com:4569/pmo
IAX-URI.4.      iax:pm2vpmo.orange-ftgroup.com:4569/pMo
IAX-URI.5.      iax:21.56.36.36:4569/pMo
IAX-URI.6.      iax:21.56.36.36/pMo
```

DNS system by **AAAA** records, while IPv4 addresses are represented as **A** records). A set of examples of equivalent IAX URIs is provided in Table 3.9.

In these examples, we suppose that the **pm2vpmo.orange-ftgroup.com** DNS resolving address is **21.56.36.36**. These URIs are equivalent since their **host** parts all refer to the same IP address and are bound to the same port number, **4569**. The **name** part is not case sensitive. For all these URIs, the **name** part refers to an equivalent value, **pmo**.

3.5 IAX Uniform Resource Identifiers and ENUM

This section provides some insights regarding the use of ENUM (Telephone Number Mapping) and IAX URIs. A brief introduction to ENUM is provided. For more details about this architecture, readers are invited to refer to the 'Further Reading' section.

3.5.1 ENUM Architecture: Overview

Most voice over IP (VoIP) protocols use URIs to identify the endpoint to be called, and therefore to initiate and route a call towards that destination. Nevertheless, VoIP is not globally implemented or even deployed. By 'globally', we mean deployment at a large scale and wide adoption. VoIP gets a small portion of world traffic compared to PSTN, although it is continuously increasing and reaches important ratios in some countries, such as France (VoIP constituted 38% of total traffic in France in Q4 2007).

Given the presence of diverse telephony and voice realms, E.164 [E.164] number schemes should coexist with SIP and IAX URIs ones. From this perspective, a method must be defined and deployed to ease the interconnection of these two worlds (E.164 and VoIP) and convert E.164 numbers into routable URIs inside VoIP realms, according to adopted procedures such as DNS-based ones. As an outcome, ENUM (Telephone Number Mapping), an architecture that allows the transformation of E.164 numbers into DNS-compatible names, has been introduced, defined and promoted within the IETF (Internet Engineering Task Force), particularly within the ENUM Working Group. This working group has produced several RFCs (Requests for Comments), such as [ENUM].

ENUM is used to ease convergence of services and to associate user identifications with a single DNS record. These services are advertised and can be discovered through questioning DNS service. A dedicated resource is defined: **e164.arpa**. This is populated, in a distributed and fully decentralized manner, to provide the infrastructure in DNS for storage of E.164 numbers. Figure 3.1 provides an overview of a tiered ENUM architecture as defined by the ITU (International Telecommunication Union).

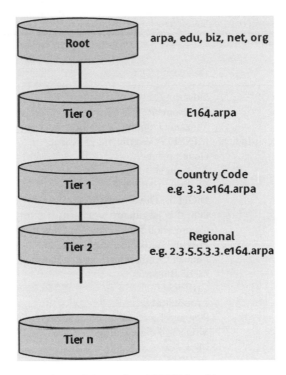

Figure 3.1 A tiered ENUM architecture

A set of E.164 country codes is given in Table 3.10. The ITU-T TSB (International Telecommunication Union, Telecommunication Standardization Sector, Telecommunication Standardization Bureau) has issued approvals for ENUM delegations to be performed on these by RIPE NCC (Network Coordination Centre).

Not all of the countries listed in Table 3.10 have opted for a public deployment of ENUM. Countries which have tried a public ENUM deployment include the USA, UK, France, Poland, Japan and Germany. These trials have shown a risk to privacy and confidentiality and a lack of robust security methods, but also the potential to ease invocation of heterogeneous services. The challenge for the Internet community now is to implement a viable and operational ENUM service which meets legal requirements and eases deployment of innovative services.

Beside the public deployment model, a private one may be investigated. Indeed, a private use of ENUM has been identified and taken into consideration to implement interconnection of service providers' realms. This model is mainly promoted within the IETF through an initiative called SPEERMINT (Session PEERing for Multimedia INTerconnect) in the context of the concept of 'federations'. This concept aims to group a set of service providers, which can share their telephony routing information and therefore ease the placement of interdomain telephony services.

This federation-based approach is not scalable and cannot meet all service providers' requirements, such as optimisation of perceived quality of service (QoS), optimisation of

Table 3.10   List of public ENUM delegations as approved by RIPE NCC

E.164 Country Code	Country	Delegee
246	Diego Garcia	Government
247	Ascension	Government
255	Tanzania	Tanzania Communications Regulatory Authority
262	France de l'Océan Indien	MINEFI (Government of France)
290	Saint Helena	Government
30	Greece	Telecom and Post Commission
31	Netherlands	Stichting ENUM Nederland
33	France	DiGITIP (Government)
350	Gibraltar	Gibraltar Regulatory Authority
353	Ireland	Commission for Communications Regulation
354	Iceland	Post and Telecom Administration
358	Finland	Finnish Communications Regulatory Authority
359	Bulgaria	ISOC Bulgaria
36	Hungary	CHIP/ISzT
374	Armenia	Arminco Ltd
386	Slovenia	Post and Electronic Communications Agency
39	Italy	Ministerio delle Communicazioni
40	Romania	MinCom
41	Switzerland	OFCOM
420	Czech Republic	Ministry of Informatics
421	Slovak Republic	Ministry of Transport, Post, and Telecommunications
423	Liechtenstein	SWITCH
43	Austria	Regulator
44	UK	DTI/Nominum
46	Sweden	NPTA
47	Norway	Norwegian Post and Telecommunications Authority
48	Poland	NASK
49	Germany	DENIC
508	Saint-Pierre et Miquelon	MINEFI (Government of France)
55	Brazil	Brazilian Internet Registry
590	Guadeloupe	MINEFI (Government of France)
594	Guyane	MINEFI (Government of France)
596	Martinique	MINEFI (Government of France)
61	Australia	Dept of Communications, Information Technology and the Arts
62	Indonesia	Directorate General of Posts and Telecommunications
63	Philippines	Commission on Information and Communications Technology
65	Singapore	IDA (Government)
66	Thailand	CAT TELECOM
81	Japan	Ministry of Internal Affairs and Communications
82	Korea (Republic of)	NIDA (National Internet Development Agency of Korea)
84	Vietnam	Ministry of Posts and Telematics of Vietnam
86	China	CNNIC
878 10		VISIONng
971	United Arab Emirates	Etisalat
882 34		Global Networks Switzerland AG

interconnection costs, implementation of sophisticated load balancing, avoidance of congestion phenomena on the path, control of both end-to-end media and signalling paths and so on.

Note that when ENUM is deployed in a private context, **e164.arpa** must not be used as a suffix for the dialing plan. This is mainly to avoid confusion with specifications pertaining to public deployment.

In the following sections, ENUM validation rules and IAX-embedded records are provided.

3.5.2 ENUM Validation Rules

ENUM is specific to E.164 numbers. Other schemes may exist, but an ENUM-enabled client queries DNS for what it believes to be an E.164 number. A critical step in the deployment of ENUM architectures is the provisioning of ENUM entries. To guide this operation, ENUM specifications define a set of rules in order to convert an E.164 number into a valid DNS record and ensure interoperable implementations.

Concretely, the following steps must be followed and executed:

- Step 1: Remove all characters except the + one and the digits. This operation is commonly referred to as 'Application Unique String' (AUS).
- Step 2: Remove the + character from the AUS.
- Step 3: Put dots (.) between the digits.
- Step 4: Reverse the order of the digits.
- Step 5: Append the string **.e164.arpa** to the end. The resulting domain name is used to request NAPTR (Naming Authority Pointer, [RFC3401]) records, which may contain either the end result or new keys in the form of DNS names to be used in further queries.

In order to illustrate this process, let us consider its output if applied to this E.164 number: **+33-2-31-75-92-31**.

- Output of Step 1: **+33231759231**.
- Output of Step 2: **33231759231**.
- Output of Step 3: **3.3.2.3.1.7.5.9.2.3.1**.
- Output of Step 4: **1.3.2.9.5.7.1.3.2.3.3**.
- Output of Step 5: **1.3.2.9.5.7.1.3.2.3.3.e164.arpa**.

3.5.3 Call Routing in an ENUM-Enabled Telephony Domain

ENUM should not be confused with an IP telephony routing protocol. ENUM is simply a protocol that eases interconnection of telephony realms. Other protocols such as TRIP (Telephony Routing over IP, [TRIP]) should be considered for end-to-end telephony placement. Other constraints, such as optimising end-to-end cost and enhancing experienced QoS, are not addressed by ENUM. These issues are out of the scope of this chapter.

This section aims to provide an overview of the use of ENUM in interconnecting telephony realms. Figure 3.2 provides an example of the use of ENUM in placing a call toward a destination identified by an E.164 number. This destination is attached to an SIP-based telephony service provider's domain.

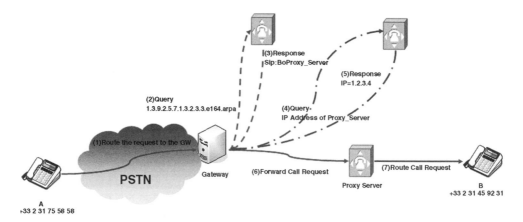

Figure 3.2   Using ENUM to place a call

In this example, a SIP-enabled service is offered, owing to a deployment of a server denoted as **Proxy Server**. The interconnection between the PSTN and SIP realms is implemented using a dedicated node called **Gateway**. We assume that ENUM service is invoked by the interconnection node itself and not by the SIP proxy server. A DNS server, which is used to retrieve the IP addresses of SIP servers, is also deployed. Only a macroscopic view of the SIP service platform is shown (a single proxy server). Note that this SIP-based realm may be deployed using IMS (IP Multimedia Subsystem, [IMS]) or other architectures.

In order to establish a call between **A** (identified by the telephony number + **33231755858**) and **B** (identified by the telephony number + **33231759231**), the following steps are observed:

- First **A** dials **B**'s number. This request is routed in the PSTN network to dedicated equipment called **Gateway**, which is an interconnection node between the PSTN and VoIP realms. This node implements interworking functions.
- Once **Gateway** receives the call request, it requests its attached ENUM server to retrieve the next hop to which the request should be relayed.
- In response, **Gateway** receives an SIP URI.
- In order to forward the request, **Gateway** needs to retrieve the IP address of the **Proxy Server**. A DNS resolution request is issued.
- In response, an IP address is sent to **Gateway**. Once it is received, the call request is forwarded to the **Proxy Server**, which in turn forwards it to **B**.

3.5.4 ENUM Service Registration

ENUM defines 'E2U' for NAPTR service. 'E2U' is used to denote an ENUM-only record. 'E2U' must be followed by one or more ENUM services. More precisely, the format of an ENUM-compatible NATPR is given in Table 3.11 (for more details about the format of a NAPT RR, refer to [RFC2915]).

Some details regarding the meaning of the tags enclosed in Table 3.11 are given below:

- **label**: specifies the domain-name format of a given E.164 number.

Table 3.11 ENUM NAPTR

label IN NAPTR order preference "u" "E2U + enumservice" regexp

- **order:** an unsigned integer specifying the order in which the NAPTR record is to be processed so as to ensure a correct ordering of the defined rules. Records with low **order** values are processed before the ones with high values. The length of this tag is **16 bits**.
- **preference:** an unsigned integer specifying the order in which NAPTR records with equal **order** values are to be processed. Records with low **preference** values are processed before those with high values. In this context, **preference** is used to compare rules that are considered the same from an authority standpoint but not from a simple load-balancing standpoint. The length of this tag is **16 bits**.
- **u:** this flag means that the next step in the DNS resolution process is not a DNS lookup but a final response which is an URI.
- **enumservice:** specifies what type of URI can be used. Examples of **enumservice** values are: SIP [RFC 3764], 'tel' [RFC4415], HTTP [RFC4002], XMPP [RFC4979], e-mail [RFC4355], fax [RFC4355], SMS [RFC4355], MMS [RFC4355] and so on.
- **regexp:** this field is a string. It encloses a replacement expression that is applied to the original string held by the client in order to construct the next domain name to be looked up. In the context of ENUM, **regexp** specifies how the AUS must be replaced.

ENUM specifications mandate that any **enumservice** is to be registered within the IANA (Internet Assigned Numbers Authority). These registrations are available at www.iana.org/assignments/enum-services. Table 3.12 provides an excerpt from this ENUM registry,

3.5.5 IAX ENUM Service Registration

As far as IAX is concerned, it is recommended to provision NAPTR resource records with appropriate IAX URIs [RFC4395]. The IAX ENUM service registration is defined in [IAXENUM]. The IAX ENUM service indicates that the resources identified by a given IAX URI are IAX capable and can be contacted using IAX protocol. When an IAX related NAPTR is returned by the ENUM service, the requesting party needs to support IAX protocol and its associated objects.

3.5.6 Examples of IAX ENUM Service Registration

In this section a set of examples illustrating records enclosing IAX URIs is provided.

3.5.6.1 Basic Registration

Table 3.13 gives an example of an NAPTR.

For more information about the meaning of the enclosed flags, readers are invited to refer to [RFC2915]. The presence of **E2U + iax** in the NAPTR record indicates that the resource **1.3.2.9.5.7.1.3.2.3.3.e164.arpa** is reachable through IAX protocol version 2 at **21.56.36.36**. The **u** flag means that this is a final response. The regexp part of this record encloses particularly **!^.*$!** and **i** at the end of the record. The meaning of these symbols in the context of an NAPTR is as follows:

Table 3.12   Excerpt from ENUM service registry

Service Name: "H323"
    URI Scheme(s): "h323:"
    Functional Specification: See Section "3. The E2U + H323 ENUM Service" of
[RFC3762]
    Security considerations: see section "5. Security Considerations" of
[RFC3762]
    Intended usage: COMMON
    Author: Orit Levin
    [RFC3762]

Service Name: "SIP"
    Type(s): "SIP"
    Subtype(s): N/A
    URI Scheme(s): "sip", "sips:"
    Functional Specification: see Section 4 of [RFC3764]
    Security considerations: see Section 6 of [RFC3764]
    Intended usage: COMMON
    Author: Jon Peterson (jon.peterson&neustar.biz)
    Any other information that the author deems interesting: see Section 3
      of [RFC3764]
    [RFC3764]

Service Name: "ifax"
    Type: "ifax"
    Subtype: "mailto"
    URI Scheme: "mailto"
    The URI Scheme is "mailto" because facsimile is a profile of
    standard Internet mail and uses standard Internet mail addressing.
    Functional Specification: see section 1 of [RFC4143]
    Security Considerations: see section 3 of [RFC4143]
    Intended usage: COMMON
    Author: Kiyoshi Toyoda (toyoda.kiyoshi&jp.panasonic.com)
            Dave Crocker (dcrocker&brandenburg.com)
    [RFC4143]

Service Name: "pres"
    URI Scheme(s): "pres:"
    Functional Specification: see Section 4 of [RFC3953]
    Security considerations: see Section 6 of [RFC3953]
    Intended usage: COMMON
    Author: Jon Peterson (jon.peterson&neustar.biz)
    Any other information that the author deems interesting: See Section 3
      of [RFC3953]
    [RFC3953]

Table 3.13   Example of IAX NAPTR configuration

$ORIGIN 1.3.2.9.5.7.1.3.2.3.3.e164.arpa.
    NAPTR 10 100 "u" "E2U + iax:iax" "!^.*$!iax:21.56.36.36/pMo!i".

- The exclamation mark (!) is used as a delimiter. It specifies the replacement output.
- **i** is used to indicate that case should be ignored.
- ^.*$ is used to indicate that all characters from the beginning to the end should be replaced.

When applied, this regular expression means that the resulting URI is **iax:21.56.36.36/pmo**.

3.5.6.2 Multiple Registrations

ENUM specifications allow us to register several entries for the same resource. To illustrate this capability, Table 3.14 shows the NAPTR records associated with **1.3.2.9.5.7.1.3.2.3.3.e164. arpa**. Six records have been entered, each of which is related to a given protocol and/or service.

Table 3.14 Examples of multiple ENUM registration

```
$ORIGIN 1.3.2.9.5.7.1.3.2.3.3.e164.arpa.
    NAPTR 10 100 "u" "E2U+SIP"           "!^.*$!sip:pmo@orange.com!".
    NAPTR 10 101 "u" "E2U+H323"          "!^.*$!h323:pmo@orange.com!".
    NAPTR 10 102 "u" "E2U+iax:iax"       "!^.*$!iax:pmo@orange.com!".
    NAPTR 10 104 "u" "E2U+tel"           "!^.*$!tel:+33-2-31-75-58-58!".
    NAPTR 10 103 "u" "E2U+email:mailto"  "!^.*$!mailto:pmo@orange.com!".
    NAPTR 10 105 "u" "E2U+http"          "!^.*$!http://www.orange.fr!".
```

1.3.2.9.5.7.1.3.2.3.3.e164.arpa is preferably reachable using (in this order) SIP, H.323, IAX, PSTN, e-mail and, lastly, HTTP. Concretely:

- If SIP is supported then the resulting URI is **sip:pmo@orange.com**. Else if SIP is not supported, go to next step:
- If H.323 is supported then the resulting URI **h323:pmo@orange.com**. Else if H.323 is not supported, go to next step:
- If IAX is supported then the resulting URI is **iax:pmo@orange.com**. Else if IAX is not supported, go to next step:
- If 'tel' scheme is supported then the resulting URI is **tel:+33-2-31-75-58-58**. Else if 'tel' is not supported, go to next step:
- If SMTP is supported then the resulting URI is **mailto:pmo@orange.com**. Else if SMTP is not supported, go to next step:
- A HTTP URL is returned: **http://www.orange.fr.**

The appropriate means to resolve protocol-specific URIs are to be invoked once a given URI is retrieved from the ENUM system.

3.5.6.3 IPv6-Embedded Registration

As stated above, IPv6 may be used to store an NAPTR record in ENUM registries. For illustration purposes, Table 3.15 shows the use of the ENUM service with an associated NAPTR

Table 3.15 Example of ENUM registration embedding an IPv6 address

```
$ORIGIN 1.3.2.9.5.7.1.3.2.3.3.e164.arpa.
    IN NAPTR 10 100 "u" "E2U+iax:iax" "!^.*$!iax:[2001:688::1]:4569/pmo!".
```

resource record that contains an IPv6 destination address. This contact information indicates that the domain **1.3.2.9.5.7.1.3.2.3.3.e164.arpa** may be contacted by using the IAX protocol at IPv6 address **2001:688::1**, port **4569**, with called party **pmo**. Note that in this record, the IPv6 address is enclosed between brackets so as to avoid confusion with ':', used to separate an IPv4 address and a port number. These brackets are not part of the IPv6 address itself.

3.5.6.4 Registration with a Context

IAX records in ENUM may also enclose a context, which characterises the invocation context in which an IAX resource may be reached.

Table 3.16 provides an example of an NAPTR record of a resource identified by **1.3.2.9.5.7.1.3.2.3.3.e164.arpa**, reachable using IAX at **21.56.36.36** in the context of **Friend**.

Table 3.16 Example of IAX ENUM registration with a context (1)

```
$ORIGIN 1.3.2.9.5.7.1.3.2.3.3.e164.arpa.
   NAPTR 10 100 "u" "E2U+iax:iax" "!^.*$!iax:21.56.36.36/pMo?Friend!".
```

A filter must be applied to return only the record which matches the invocation context. Therefore, if an ENUM record encloses the following lines and the requestor asks to reach the IAX resource which belongs to the **Friend** context, the returned IP address must be **21.56.36.36** and not **21.56.15.23**, as shown in Table 3.17.

Table 3.17 Examples of IAX ENUM registration with a context (2)

```
$ORIGIN 1.3.2.9.5.7.1.3.2.3.3.e164.arpa.
   NAPTR 10 100 "u" "E2U+iax:iax"   "!^.*$!iax:21.56.36.36/pMo?Friend!".
   NAPTR 10 100 "u" "E2U+iax:iax"   "!^.*$!iax:21.56.15.23/pMo?PRO!".
```

References

[E164] ITU-T Recommendation E.164, 'The International Public Telecommunication Number Plan', May 1997.

[ENUM] Faltstrom, P. and Mealling, M., 'The E.164 to Uniform Resource Identifiers (URI) Dynamic Delegation Discovery System (DDDS) Application (ENUM)', RFC 3761, April 2004.

[H.323] ITU-T Recommendation H.323, 'Packet-Based Multimedia Communications Systems', International Telecommunication Union (ITU-T), November 2000.

[IAX] Spencer, M., Shumard, K., Capouch, B. and Guy, E.,'IAX2: Inter-Asterisk eXchange Version 2', draft-guy-iax-04, work in progress.

[IAXENUM] Guy, E., 'IANA Registration for IAX Enumservice', draft-ietf-enum-iax, work in progress.

[IMS] Camarillo, G. and Garcia-Martin, M.A., 'The 3G IP Multimedia Subsystem: Merging the Internet and the Cellular Worlds', John Wiley and Sons, Ltd., 2005.

[MGCP] Andreasen, F. and Foster, B., 'Media Gateway Control Protocol (MGCP) version 1.0', RFC3435, January 2003.

[RFC2915] Mealling, M. and Daniel, R., 'The Naming Authority Pointer (NAPTR) DNS Resource Record', RFC 2915, September 2000.

[RFC3401] Mealling, M., 'Dynamic Delegation Discovery System (DDDS) Part One: The Comprehensive DDDS', RFC 3401, October 2002.

[RFC3764] Peterson, J., 'Enumservice Registration for SIP Addresses-of-Record', RFC 3764, April 2004.

[RFC3953] Peterson, J., 'Enumservice Registration for Presence Services', RFC 3953, January 2005.

[RFC3986] Berners-Lee, T., Fielding, R. and Masinter, L., 'Uniform Resource Identifier (URI): Generic Syntax', STD 66, RFC 3986, January 2005.

[RFC4002] Brandner, R., Conroy, L. and Stastny, R., 'IANA Registration for ENUMservices Web and ft', RFC 4002, February 2005.

[RFC4143] Toyoda, K. and Crocker, D., 'IFAX Service of ENUM', RFC 4143, November 2005.

[RFC4291] Hinden, R. and Deering, S., 'IP Version 6 Addressing Architecture', RFC 4291, February 2006.

[RFC4355] Brandner, R., Conroy, L. and Stastny, R., 'IANA Registration for Enumservices email, fax, mms, ems and sms', RFC 4355, January 2006.

[RFC4395] Hansen, T., Hardie, T. and Masinter, L. 'Guidelines and Registration Procedures for New URI Schemes', BCP 115, RFC 4395, February 2006.

[RFC4415] Brandner, R., Conroy, L. and Stastny, R., 'IANA Registration for Enumservice Voice', RFC 4415, February 2006.

[RFC4769] Livingood, J. and Shockey, R. 'IANA Registration for an Enumservice Containing PSTN Signaling Information', RFC 4769, November 2006.

[RFC4979] Mayrhofer, A., 'IANA Registration for Enumservice "XMPP"', RFC 4979, August 2007.

[SIP] Rosenberg, J., Schulzrinne, H., Camarillo, G., Johnston, A., Peterson, J., Sparks, R. et al., 'SIP: Session Initiation Protocol', RFC 3261, June 2002.

[SMTP] Klensin, J., 'Simple Mail Transfer Protocol', RFC2821, April 2001.

[TRIP] Rosenberg, J., et al., 'Telephony Routing over IP (TRIP)', RFC 3219, January 2002.

[TURN] Rosenberg, J., et al., 'Traversal Using Relay NAT (TURN)', work in progress.

Further Reading

Berners-Lee, T., Fielding, R. and Masinter, L. 'Uniform Resource Identifiers (URI): Generic Syntax', RFC 2396, August 1998.

Center for Democracy and Technology, 'Enum: Mapping Telephone Numbers onto the Internet Potential Benefits with Public Policy Risks', available at: http://www.cdt.org/standards/enum/.

Faltstrom, P. and Mealling, M., 'E.164 to Uniform Resource Identifiers (URI) Dynamic Delegation Discovery System (DDDS) Application (ENUM)', RFC 3761, April 2004.

Huston, G. 'ENUM: Mapping the E.164 Number Space into the DNS', *Internet Protocol Journal*, June 2002.

IETF Working Group, 'Session PEERing for Multimedia INTerconnect (speermint)', http://www.ietf.org/html.charters/speermint-charter.html.

IETF Working Group, 'Telephone Number Mapping (enum)', http://www.ietf.org/html.charters/enum-charter.html.

Levin, O., 'ENUM Service Registration for H.323 URL', RFC 3762, April 2004.

Livingood, J. and Troshynski, D., 'IANA Registration of Enumservices for Voice and Video Messaging', RFC 5278, July 2008.

Mahy, R., 'A Telephone Number Mapping (ENUM) Service Registration for Instant Messaging (IM) Services', RFC 5028, October 2007.

Mahy, R., 'A Telephone Number Mapping (ENUM) Service Registration for Internet Calendaring Services', draft-ietf-enum-calendar-service-04, work in progress, March 2008.

Mayrhofer, A., 'IANA Registration for vCard Enumservice', RFC 4969, August 2007.

Mayrhofer, A., 'IANA Registration for Location ('loc') Enumservice', draft-mayrhofer-enum-loc-enumservice, work in progress.

Mealling, M., 'Dynamic Delegation Discovery System (DDDS) Part Two: The Algorithm', RFC 3402, October 2002.

Mealling, M., 'Dynamic Delegation Discovery System (DDDS) Part Three: The Domain Name System (DNS) Database', RFC 3403, October 2002.

Mealling, M., 'Dynamic Delegation Discovery System (DDDS) Part Four: The Uniform Resource Identifiers (URI) Resolution Application', RFC 3404, October 2002.

Mealling, M. and Daniel, R., 'The Naming Authority Pointer (NAPTR) DNS Resource Record', RFC 2915, September 2000.

Mockapetris, P., 'Domain Names - Concepts and Facilities', STD13, RFC 1034, November 1987.

Peterson, J., 'Enumservice Registration for SIP Addresses-of-Record', RFC 3764, April 2004.

Vaha-Sipila, A., 'URLs for Telephone Calls', RFC 2806, April 2000.

Vaudreuil, G., 'Voice Message Routing Service', RFC 4238, October 2005.

4

IAX Frames

4.1 Introduction

In order to ease your understanding of information conveyed in IAX messages, which is used to determine IAX behaviour with regards to a given event (for example receipt of an IAX message), we have decided to present the structure of IAX messages and their taxonomy before describing IAX operations (see Chapter 8). This chapter focuses on the structure of IAX frames, while Chapter 5 is dedicated to IAX informational elements (IE).

IAX distinguishes between several types of frame. Three major categories are manipulated by the IAX protocol: 'mini', 'full' and 'meta' frames. IAX also conveys encrypted frames, but these are not considered an IAX frame type per se.

Table 4.1 illustrates the three main categories of IAX frame and the difference between them in terms of their uses and structures. For each frame type the following information is provided:

- *Main Characteristic*: identifies its main feature, and what makes it different from the others.
- *Header Size*: provides its header size.
- *Use*: gives a brief description of its main uses.

The rest of the chapter provides a detailed description of each type of IAX frame.

4.2 Full Frames

4.2.1 Structure

An IAX full frame is generally used to convey signalling messages. Media data may also be enclosed in a full frame, but this is not recommended and dedicated frames should be used to carry media data.

IAX full frames are used for reliable message exchange. An acknowledgement message must be sent by the remote IAX participant to note that the full frame has been received and/or processed. This acknowledgement message may be:

Table 4.1 IAX frames

Frame Type	Main Characteristic	Header Size	Use
Mini	This frame does not require any acknowledgement from the remote IAX participant	4 bytes	This type of frame is usually used to convey voice data or other media data
Full	This frame requires an acknowledgement from the remote IAX participant	12 bytes	This type of frame is used to send reliable data, such as IAX control messages. Media data may also be carried in this type of frame
Meta	This frame is unreliable	6 bytes for video meta frames and 8 bytes for meta trunk frames	This type of frame is used to convey video, or multiple mini frames with a single IAX header

- A protocol-specific (more precisely, context-specific) acknowledgment such as a **REGACK** message (see Chapter 6 for more details about this message).
- An explicit acknowledgment, thanks to the use of an **ACK** message (see Chapter 6 for more details about this message).

Figure 4.1 shows the structure of a full frame. The length of a full frame header is **12 bytes**. The following fields must be included when sending an IAX full frame:

- **F**: this bit must be set to **1** to indicate that this is a full frame.
- **Source Call Number**: this field stores an identifier assigned by the local IAX speaker to unambiguously identify the call among all existent active ones (maintained by this IAX endpoint). This identifier may be reused within the context of another call if this call context has been destroyed.

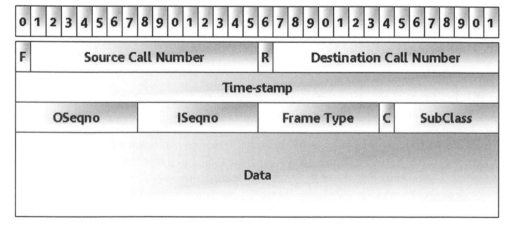

Figure 4.1 Full frame format

The length of this field is **15 bits**.
- **R**: this bit notifies the remote peer if this is a retransmitted frame:

  - if set to **1**, this is a retransmitted frame
  - if set to **0**, this frame is beingsent for the first time.

- **Destination Call Number**: this field stores the identifier assigned by the remote IAX peer to identify a given call on its side. This number has to be set to the same value as the one enclosed as **Source Call Number**, conveyed in IAX frames issued by the remote IAX peer within the same call.
  The length of this field is **15 bits**.
- **Time-stamp**: this field is used to convey an incremental representation of the number of milliseconds since the first transmission within the context of this call.
  The length of this field is **32 bits**.
- **OSeqno**: the purpose of this field is to identify the outbound streams within a given call. This is achieved by maintaining an outbound sequence number per call. The outbound sequence number enclosed in the first issued message within a call is set to **0**.
  The length of this field is **8 bits**.
- ISeqno: the purpose of this field is to identify the received streams within a given call. An inbound sequence number per call is to be maintained by an IAX speaker. This identifier value is to be set to **0** upon call initialisation.
  The length of this field is **8 bits**.
- **Frame Type and SubClass**: these fields indicate the frame type (**8 bits**) and subclass (**7 bits**) of the message carried by the frame. These fields may take the values indicated in Table 4.2.
- **C**: this field is used to indicate the format of the **SubClass** field:

  - if set to **0**, the **SubClass** field value is interpreted as an unsigned integer
  - if set to **1**, the **SubClass** field value is interpreted as power of **2**.

- **Data**: this field carries the data. Its content depends on the frame type.

Table 4.2 gives a set of frame types supported by IAX protocol. For each frame type, the following data is provided:

- *Full Frame Type*: provides the IAX code used to unambiguously identify the type of full frame under consideration. Only this value is carried in the IAX frames exchanged between IAX peers.
- *Full Frame Name*: indicates the name used to refer to the type of full frame under consideration. In the remainder of this book, IAX full frames are referred to by their short name. Therefore, a voice frame should be understood as 'IAX full frame of voice type', IAX control messages should be understood as 'IAX control full frames' and so on.
- *Description*: provides more detail regarding the full frame type and its invocation context.
- *Subclass*: one or several subclasses may be associated with each full frame type. This column provides a set of associated subclasses defined by IAX specifications for each full frame type. Additional subclasses may be defined and supported by IAX implementation. The list of subclasses for each full frame type is not exhaustive and those provided are for illustration purposes only.

Table 4.2 Frame types and subclasses

Full Frame Type	Full Frame Name	Description	Subclass
0x01	DTMF	This type is used to carry a single digit of DTMF (Dual Tone Multiple Frequency, [DTMF])	DTMF digit (i.e. 0-9, A-D, *, Full Frame Type, #)
0x02	Voice	This type is used to carry voice data	The audio subclass value indicates a valid audio format, such as G.711 [G711] or iLBC [ILBC]
0x03	Video	This type is used to carry video data	The audio subclass value indicates a valid video format, such as H.263 [H263] or H.261 [H261]
0x04	Control	This type is used to carry information about the status of an ongoing call session	Examples of subclass values are listed below: • 0x03: Ringing • 0x04: Answer • 0x05: Busy • 0x08: Congestion • 0x0b: Option • 0x0e: Call Progress • 0x0f: Call Proceeding
0x05	Null	This type should never be sent	NA
0x06	IAX2	This type is used to provide endpoint call management	Examples of subclass values are listed below (a full list is given in Chapter 6): • 0x01: NEW • 0x02: PING • 0x03: PONG • 0x04: ACK • 0x06: REJECT • 0x08: AUTHREQ • 0x25: FWDATA
0x07	Text	This type is used to carry a text IAX message	The text subclass value is 0
0x08	Image	This type is used to convey a single image	The subclass indicates a valid image format, such as JPEG (Joint Photographic Experts Group)
0x09	HTML	This type is used to convey HTML data	The accepted subclass values are listed below: • 1: Sending URL • 2: Data frame

Table 4.2 (*Continued*)

Full Frame Type	Full Frame Name	Description	Subclass
			• 4: Beginning frame
			• 8: End frame
			• 16: Load is complete
			• 17: The peer does not support HTML
			• 18: Link URL
			• 19: Unlink URL
			• 20: Reject link URL
			For more information about the exploitation of these values, please refer to [IAX]
0x0A	Comfort Noise	This type is used to convey comfort noise	The subclass value indicates the level of comfort noise in dBov[1]

[1]dBov is the level relative to the overload of a system. For more information, please refer to [RTP].

4.2.2 Trace Example

Figure 4.2 illustrates a trace example of an IAX full frame. As illustrated, this frame is a full frame since the **F** bit is positioned to **1**. In this example, the **Destination Call Number** is positioned to **0**, because this is the first message issued by the given IAX peer. Only the **Source Call Number** is positioned to **16384**. This full frame is an IAX control message, since the value of the **Type** field is positioned to **6**, and the **SubClass** is **NEW** since the value of this field is **1**.

4.3 Mini Frames

4.3.1 Structure

Because the header length is reduced to **4 octets**, this type of IAX frame is called a mini frame. Mini frames are used only to send voice flows once a call session has been established between two IAX peers. Unlike full frames, mini frames do not require any acknowledgement from the remote participant (that is, these messages are unreliable).

IAX mini frames are assumed to carry exclusively voice and/or video data, encoded according to the CODEC (Compression/Decompression) negotiated during the call initiation phase. The CODEC used to exchange voice streams may be modified during an ongoing communication/session. To do so, a full frame (with a new value in its **SubClass**) field must be issued by one of the communication participants. After receiving this full frame, further mini frames should be issued according to the new CODEC indicated in the latest received IAX full frame.

The structure of an IAX mini frame is sketched in Figure 4.3.

The following fields must be enclosed when sending an IAX mini frame:

- **F**: this bit must be set to **0** to indicate that this is not a full frame. Once this bit is set to **0**, the frame is considered a mini frame.

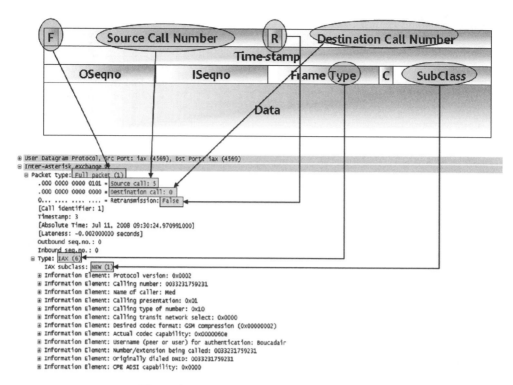

Figure 4.2   Example of an IAX full frame trace

- **Source Call Number**: this field stores an identifier assigned by a local IAX speaker to
  unambiguously identify the call among all existent active ones. This identifier may be
  reused within the context of another call if this call context has been destroyed and is no
  longer valid.

  The length of this field is **15 bits**.
- **Time-stamp**: this field carries the lower **16 bits** of the transmitting peer's full **32 bit**
  timestamp, as defined in Section 4.2.1). The timestamp value is used to (re)order the received

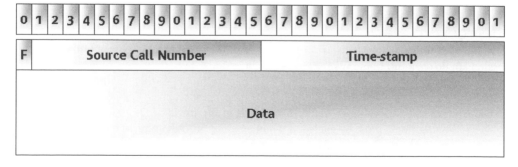

Figure 4.3   Mini frame format

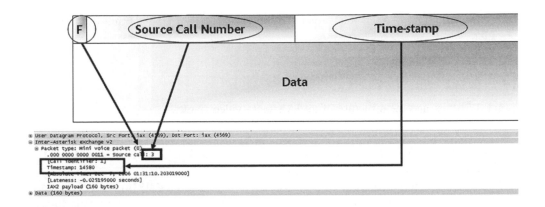

Figure 4.4   Example of a mini frame trace

frames. If the value of this field reaches **65536** seconds, the following mini frames will have a
timestamp value set to **0**.

The length of this field is **16 bits**.

- **Data**: this field carries the data. The length of the enclosed data may be up to the maximum
value supported by the network.

4.3.2 Trace Example

Figure 4.4 illustrates a trace example of an IAX mini frame. This frame is a mini frame because
the **F** bit is positioned to **0**. The **Source Call Number** is positioned to a non null value. To
encode this mini frame, a given IAX peer must use the CODEC conveyed in the last issued/
received full frame.

4.4 Meta Frames

An IAX meta frame is used in two scenarios:

- To exchange video streams with an optimised IAX protocol header (note that these frames are
similar to the IAX mini frames).

- To allow multiple IAX media streams to be included in a single frame with a single header.

4.4.1 Meta Video Frames

Figure 4.5 illustrates the structure of an IAX meta video frame:

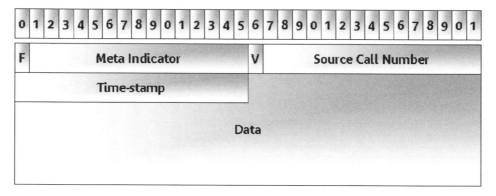

Figure 4.5   Meta video frame format

The following fields must be included when sending an IAX meta video frame:

- **F**: this bit must be set to **0** to indicate that this is not a full frame.
- **Meta Indicator**: this field is always set to zeros. It is used to identify a meta frame (thanks to the presence of **16** zeros (**F** bit included)).
  The length of this field is **15 bits**.
- **V**: this bit is set to **1** to indicate that this is a meta video frame.
- **Source Call Number**: this field carries the identifier assigned by the local IAX speaker to uniquely identify this call.
- **Time-stamp**: this field carries the lower **16 bits** of the transmitting full **32 bit** timestamp, as defined in Section 4.2.1. The timestamp value is used to (re)order the received frames. If the value of this field reaches **65 536** seconds, a full frame should be sent to the remote peer to indicate that the timestamp has set to **0**.
  The length of this field is **16 bits**.
- Data: this field carries the video-encoded data according to either the CODEC negotiated during the call setup or the video CODEC indicated in the latest received/issued full frame.

4.4.2 Meta Trunk Frames

As indicated above, the main motivation for the introduction of meta trunk frames is to optimise the amount of consumed bandwidth. This is achieved by enclosing a set of media streams exchanged between the same peers in a single message (that is, a message with a single IAX header). There are two ways to build media trunking:

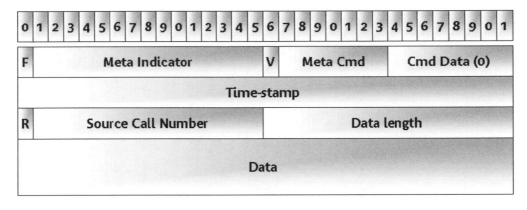

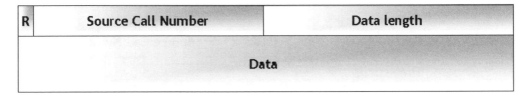

Figure 4.6   Meta trunk frame format

- The first is to send a standard meta frame header with a complete timestamp related to the trunk frame, followed by one or several media frames. Each media frame indicates a **Source Call Number** and the **Data Length** of the enclosed media stream data (see Figure 4.6).
- The second method is similar to the first, except that each media frame encloses, in addition to the **Source Call Number, Data Length** and media data, a timestamp specific to the call session (see Figure 4.7).

    The following fields must be included when sending an IAX meta trunk frame:
- **F**: this bit must be set to **0** to indicate that this is not a full frame.
- **Meta Indicator**: this field is always set to zeros. It is used to identify a meta frame (thanks to the presence of **16** zeros (**F** bit included)).

    The length of this field is **15 bits**.
- **V**: this bit is set to **0** to indicate that this is not a meta video frame.
- **Meta Command**: this field is used to indicate whether or not the meta frame is a meta trunk. Currently, the IAX specifications only accept a value equal to **1**, to indicate that this is a meta trunk frame. Other values are reserved for further use.

    The length of this field is **8 bits**.
- Command Data: this field indicates which option to apply to the trunk:

    - if the field is set to **0**, this implies that the calls do not include their timestamps
    - if the field is set to **1**, this implies that each call is associated with its timestamp.

    All remaining bits are to be set to zeros.
    The length of this field is **8 bits**.

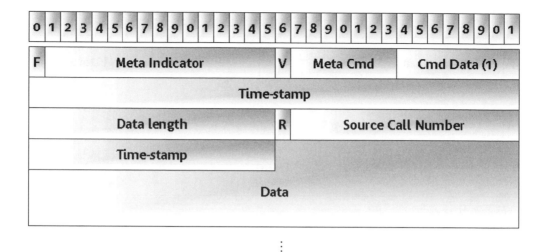

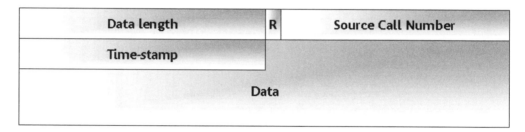

Figure 4.7   Meta trunk frame format (bis)

- Time-stamp: this field carries the timestamp related to the time of transmission of the trunk frame.

  The meanings of the remaining fields are the same as those for mini frames.

4.5 Encrypted Frames

All the aforementioned IAX frames may be encrypted. The basis specification of IAX protocol recommends the use of the AES (Advanced Encryption Standard, [AES]) procedure to cipher IAX messages. Chapter 7 provides more information about this procedure. This section does not focus on the use of AES within IAX, only on the structure of the encrypted IAX message.

Before encrypting an IAX full frame, padding is added at the front of the data because AES requires blocks with a size of **16 bytes**. Figure 4.8 illustrates the structure of a full frame after padding is added but before encryption.

The data is then encrypted using AES. Figure 4.9 illustrates the structure of an encrypted IAX full frame. As shown in this figure, the first **4 bytes** are passed in clear. The remaining parts are encrypted using the AES challenges as exchanged between two IAX peers prior to encryption. Usually the first encrypted message within a given IAX session is an **ACK** message.

The same encryption rules apply to mini frames, except that the initial unencrypted part of the frame is only **2 bytes** (**F** and **Source Call Number** fields), as illustrated in Figure 4.10.

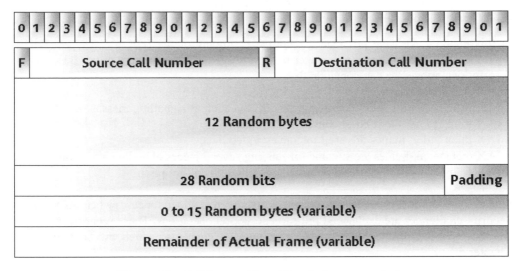

Figure 4.8   Full frame with padding (before encryption)

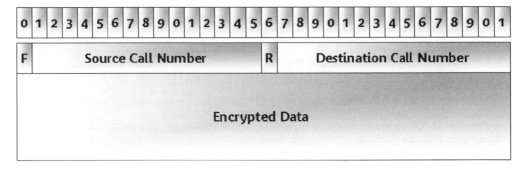

Figure 4.9   Format of an encrypted full frame

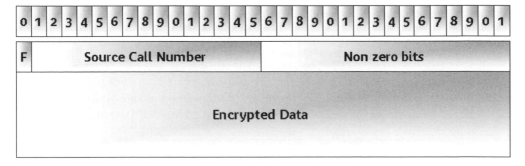

Figure 4.10   Encrypted mini frame

4.6 Conclusion

In this chapter, we have provided a tentative taxonomy of the IAX frame types. Details regarding their uses and structures (in terms of message format) have been given. Based on the descriptions above, we can conclude that:

- IAX full frames are used to convey reliable data such as signalling messages.
- IAX mini frames are used to convey voice data.
- IAX meta video frames are used to convey video data.
- IAX meta trunk frames are suitable for sending several pieces of audio data within a single IAX frame, for the purpose of optimising bandwidth consumption.

Finally, the format of IAX encrypted messages has been provided for cases where AES is used. Security challenges are exchanged prior to encryption operations. These security challenges are conveyed in dedicated objects called 'information elements', which will be described in Chapter 5.

References

[AES] US Department of Commerce/NIST, 'FIPS-197, Announcing the Advanced Encryption Standard', November 2001.
[DTMF] Schulzrinne, H. and Petrack, S.,'RTP Payload for DTMF Digits, Telephony Tones and Telephony Signals', RFC2833, May 2000.
[G711] ITU-T Recommendation G.711, 'Pulse Code Modulation (PCM) of Voice Frequencies', International Telecommunication Union (ITU-T), November 1988.
[H261] ITU-T Recommendation H.261, 'Video Codec for Audiovisual Services at px64 kbits/s', International Telecommunication Union (ITU-T), 1993.
[H263] ITU-T Recommendation H.263, 'Video Coding for Low Bit Rate Communication', International Telecommunication Union (ITU-T), July 1995.
[IAX] Spencer, M., Shumard, K., Capouch, B. and Guy, E., C 'IAX2: Inter-Asterisk eXchange Version 2', draft-guy-iax-04, work in progress.
[ILBC] Andersen, S. et al., 'Internet Low Bit Rate Codec (iLBC)', RFC 3951, December 2004.
[RTP] Schulzrinne, H., Casner, S., Frederick, R. and Jacobson, V., 'RTP: A Transport Protocol for Real-Time Applications', RFC 1889 (proposed standard), January 1996.

Further Reading

IAX Analyser, available at: http://www.unleashnetworks.com/articles/asterisk-call-analyzer-for-iax2.html.
Unleash Networks White Paper, 'IAX2 Call Analyzer for Unsniff', available at: Ethereal IAX support: http://www.unleashnetworks.com/lib/IAX2AnalyzerWhitepaper3.pdf.
Wireshark IAX Wiki, available at: http://wiki.wireshark.org/IAX2.

5

IAX Information Elements

5.1 Introduction

One of the particularities of the IAX protocol is the use of information element (IE) objects to carry information required for the management of the IAX calls. These elements are carried in IAX full frames. A list of the information elements currently used in [IAX] is given in Table 5.1. New information elements may be introduced for new usages.

The generic structure of an informational element follows the format shown in Figure 5.1.

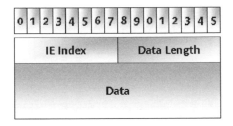

Figure 5.1 Generic information-element structure

The meanings of the enclosed fields are as follows:

- *IE Index:* stores a unique identifier which unambiguously identifies a given IE among all existing ones.
- *Data Length:* specifies the length of the enclosed data.
- *UTF8-Encoded data:* the length of this field is variable. It carries the UTF8-encoded data [UTF8] meaningful for a given IE.

Table 5.1 List of IAX information elements

Abbr. IE Name	Description	IE Message Format
Called Number CALLED NUMBER	Within IAX a called number is not limited to E.164 numbers [E164] but may also include nonnumeric characters such as an SIP URI. The use of 'number' is misleading. We recommend denoting this IE 'Called (IAX) URI' instead of 'Called Number'. This IE is used to carry the called URI	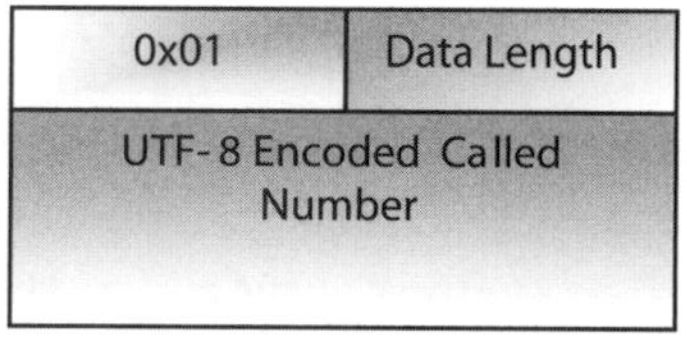 0x01 | Data Length UTF-8 Encoded Called Number **Figure 5.2** CALLED NB IE
Calling Number CALLING NUMBER	This IE is used to indicate the number/IAX URI of the calling entity	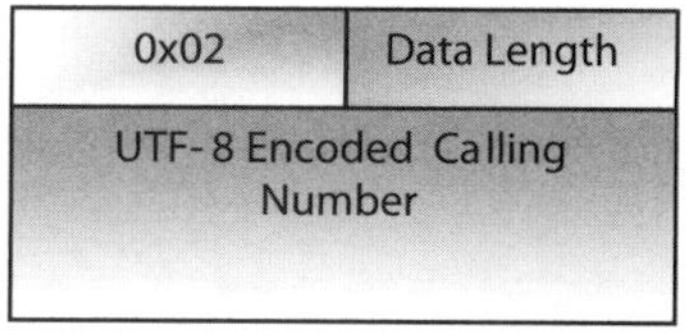 0x02 | Data Length UTF-8 Encoded Calling Number **Figure 5.3** CALLING NB IE
Calling ANI CALLING ANI	This IE carries the calling number ANI (Automatic Number Identification). The ANI is traditionally used for billing purposes[1]	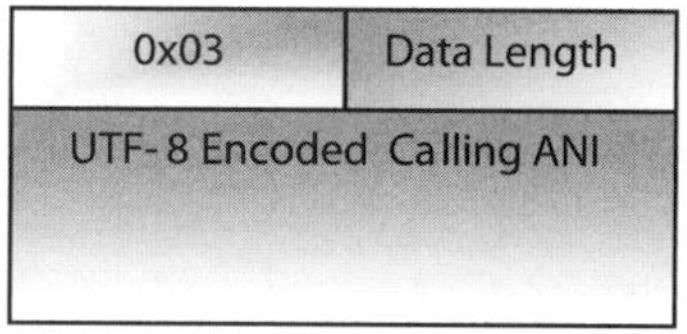 0x03 | Data Length UTF-8 Encoded Calling ANI **Figure 5.4** CALLING ANI IE

Calling Name
CALLING NAME

This IE carries the calling name of the IAX peer issuing the request

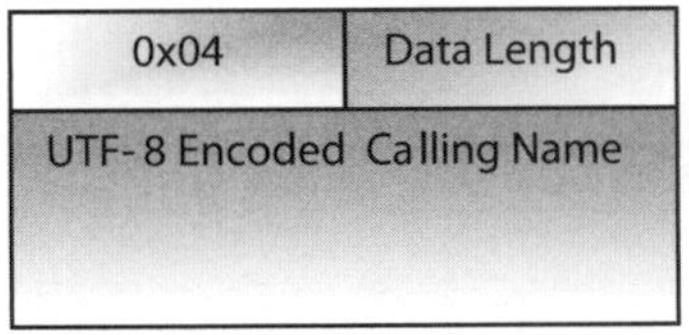

Figure 5.5 CALLING NAME IE

Called Context
CALLED CONTEXT

This IE provides an indication of the remote dial plan context of the ongoing call. Note that a context may be a line number, trunk group, etc.

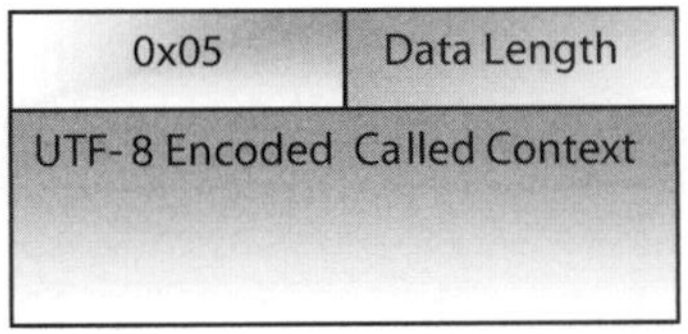

Figure 5.6 CALLED CTX IE

User Name
USERNAME

This IE carries the identity of the user issuing a given IAX message

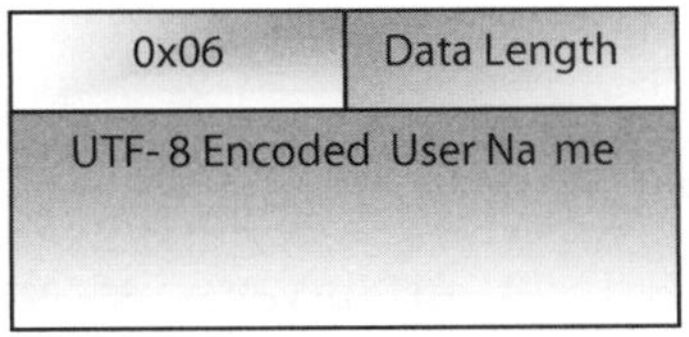

Figure 5.7 USERENAME IE

Password
PASSWORD

This IE carries a plaintext or encrypted password. The format of this IE is not specified and is application-specific

N/A

(continued)

Table 5.1 (*Continued*)

Abbr. IE Name	Description	IE Message Format

Capability
CAPABILITY

This IE is used to inform the remote IAX peer about the CODEC capabilities of the calling participant. Note that multiple CODECs may be carried in a single CAPABILITY IE.

Examples of MEDIA CAPABILITY values include (a full list of accepted CODEC values can be found at [IAX]):

- 0x00000001: G.723.1
- 0x00000002: GSM Full Rate
- 0x00000080: LPC10
- 0x00000200: SPEEX

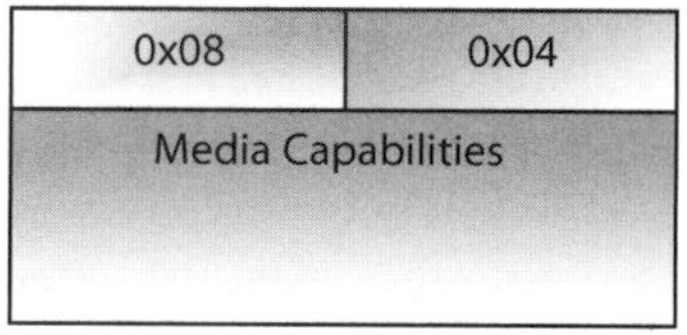

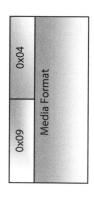

0x08	0x04
Media Capabilities	

Figure 5.8 CAPABILITY IE

Format
FORMAT

This IE is used to indicate to a remote IAX peer the single preferred media CODEC. A FORMAT IE may be enclosed in IAX messages, especially: (1) NEW messages to indicate the preferred CODEC; (2) ACCEPT messages to indicate the selected CODEC for the media exchange. Examples of MEDIA FORMAT values include (A full list of accepted CODEC values can be found at [IAX]):

- 0x00000400: ILBC
- 0x00000800: AMR
- 0x00010000: JPEG

0x09	0x04
Media Format	

Figure 5.9 FORMAT IE

For more details about NEW and ACCEPT messages, refer to Chapter 6

Language
LANGUAGE

In order to indicate the language to be used within a given call, a LANGUAGE IE may be enclosed in NEW messages. For more details about IAX NEW message, refer to Chapter 6

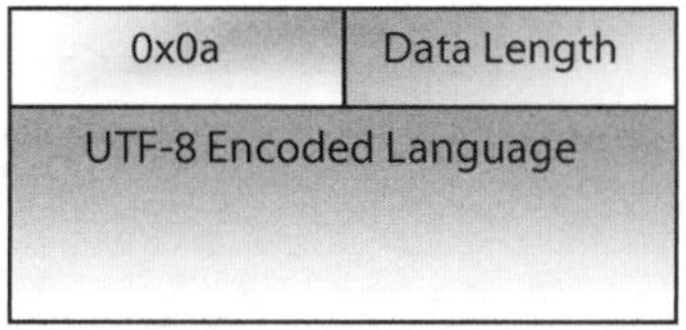

Figure 5.10 LANGUAGE IE

0x0a	Data Length
UTF-8 Encoded Language	

Version
VERSION

This IE indicates the IAX protocol version supported by a given IAX speaker. The current supported version is 2. In order for IAX call setup to be successful, the two participating peers must support the same IAX protocol version

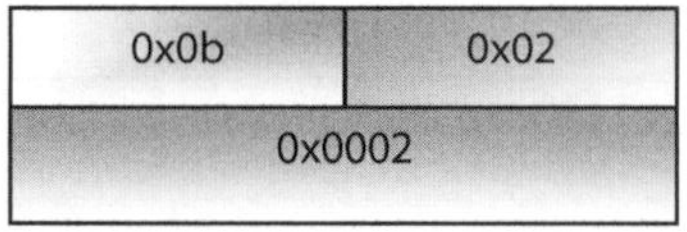

Figure 5.11 VERSION IE

0x0b	0x02
0x0002	

ADSI CPE
ADSICPE

This IE is used to indicate whether the used device has ADSI (Analog Display Services Interface) capability ADSI capability is used to pass encoded information to a handset

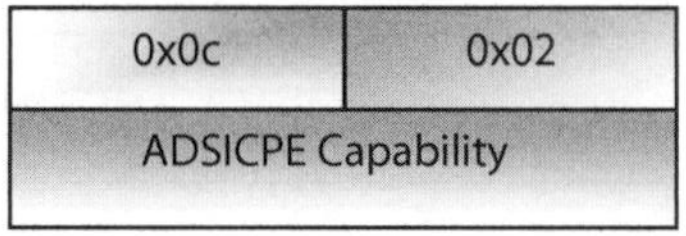

Figure 5.12 ADSICPE IE

0x0c	0x02
ADSICPE Capability	

Dialed Number ID
DNID

This IE indicates the Dialed Number ID, which may differ from the Called Number

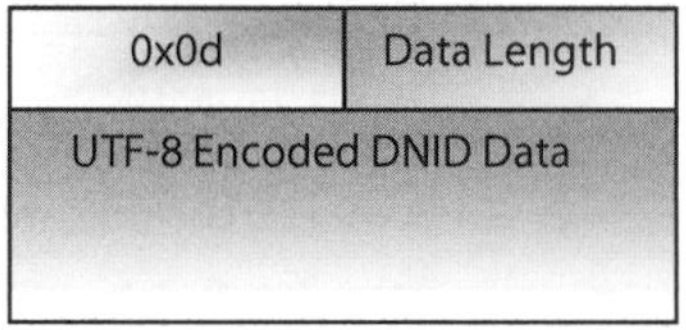

Figure 5.13 DNID IE

0x0d	Data Length
UTF-8 Encoded DNID Data	

(continued)

Table 5.1 (*Continued*)

Abbr. IE Name	Description	IE Message Format
Authentication Methods AUTHMETHODS	This IE indicates the authentication methods supported by a given IAX speaker. Valid values for authentications are: • 0x0001: Reserved • 0x0002: MD5 [MD5] • 0x0003: RSA [RSA]	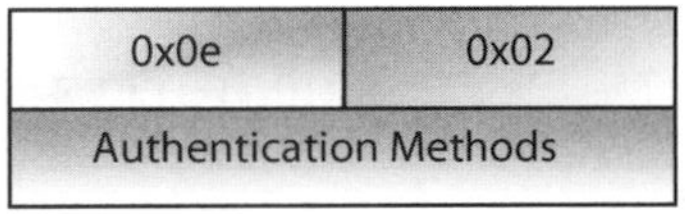 0x0e | 0x02 Authentication Methods **Figure 5.14** AUTH MDS IE
Challenge CHALLENGE	In order to challenge a remote IAX peer, this IE is used to convey an MD5 or RSA challenge, to be used for authentication purposes	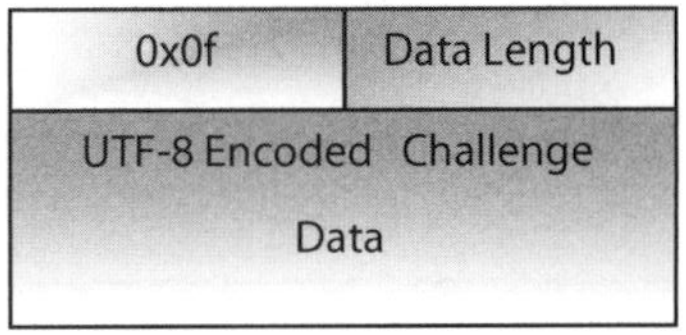 0x0f | Data Length UTF-8 Encoded Challenge Data **Figure 5.15** CHALLENGE IE
MD5 Result MD5RESULT	The result of an MD5 challenge can be conveyed in this IE	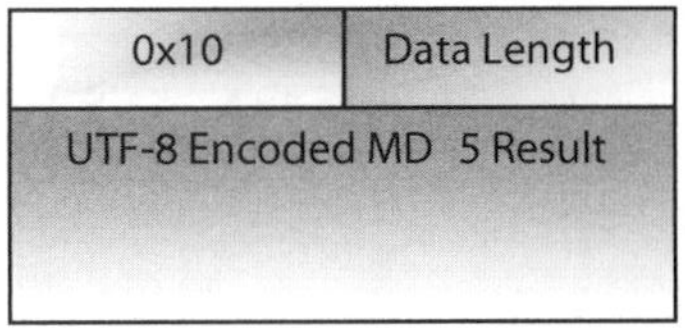 0x10 | Data Length UTF-8 Encoded MD 5 Result **Figure 5.16** MD5RESULT IE
RSA Result RSARESULT	The result of an RSA challenge can be conveyed in this IE	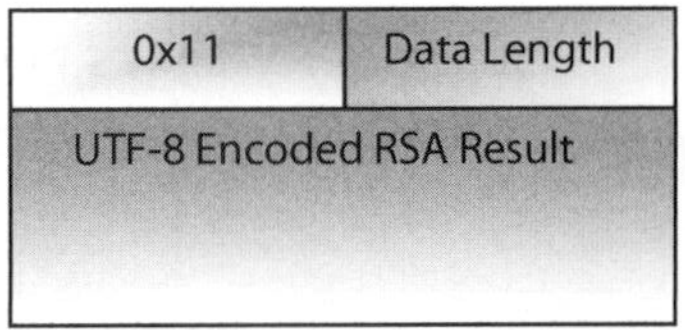 0x11 | Data Length UTF-8 Encoded RSA Result **Figure 5.17** RSARESULT IE

Apparent Address
APPARENT ADDR

This IE is used by a given IAX peer to indicate the apparent connection information (i.e. IP address and port number) of a remote IAX peer

- IPv4 Address Family

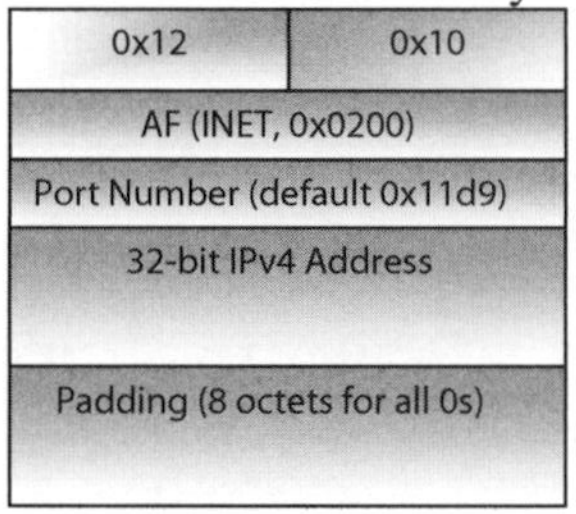

0x12	0x10
AF (INET, 0x0200)	
Port Number (default 0x11d9)	
32-bit IPv4 Address	
Padding (8 octets for all 0s)	

Figure 5.18 APPARENT IPv4 ADDR IE

- IPv6 Address Family

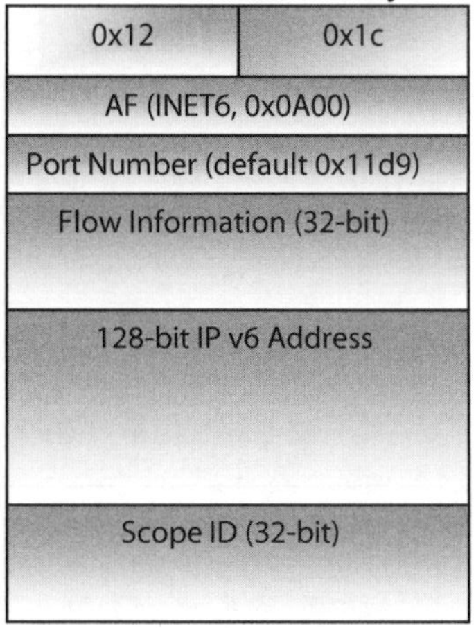

0x12	0x1c
AF (INET6, 0x0A00)	
Port Number (default 0x11d9)	
Flow Information (32-bit)	
128-bit IP v6 Address	
Scope ID (32-bit)	

Figure 5.19 APPARENT IPv6 ADDR IE

(continued)

Table 5.1 (*Continued*)

Abbr. IE Name	Description	IE Message Format
Refresh REFRESH	This IE is used by an IAX speaker to indicate the expire timer for a given event. The timer is expressed in seconds. The default value of this IE is 60 seconds	 **Figure 5.20** REFRESH IE
Dial Plan Status DPSTATUS	This IE indicates the status of a CALLED NUMBER in a remote dial plan. The 'N' bit is set to 1 if the called number is nonexistent. The 'C' bit is set to 1 to indicate that the called number can exist. 'E' is set to 1 if the called number exists. 'M' and 'R' indicate respectively 'Retain dialtone' and 'More digit may match number'	**Figure 5.21** DPSTATUS IE
Call Number CALLNO	Within a transfer operation, this IE is used to indicate the call number a remote peer needs to use to identify the ongoing call	**Figure 5.22** CALLNO IE

Cause
CAUSE

This IE is used to inform a remote peer of the reasons why an event has occurred

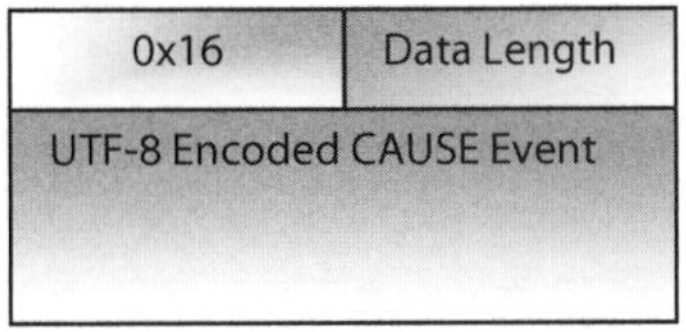

0x16	Data Length
UTF-8 Encoded CAUSE Event	

Figure 5.23   CAUSE IE

Unknown IAX2
UNKNOWN

When an unsupported IAX method has been received by an IAX peer, this IE must be issued

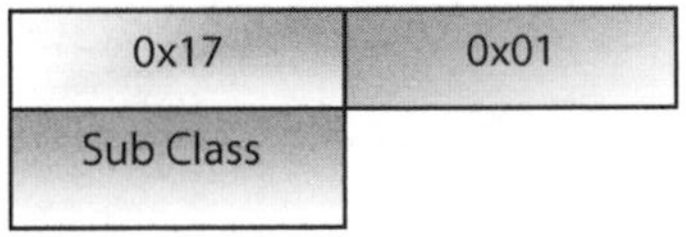

0x17	0x01
Sub Class	

Figure 5.24   UNKNOWN IE

Message Count
MSGCOUNT

This IE is used to inform a given IAX user how many voicemail messages are present in a voicemailbox, mailbox, etc. This IE indicates the number of old and new messages

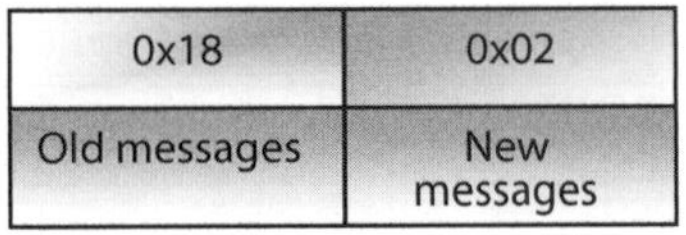

0x18	0x02
Old messages	New messages

Figure 5.25   MSGCOUNT IE

(continued)

Table 5.1 (*Continued*)

Abbr. IE Name	Description	IE Message Format
Auto Answer AUTOANSWER	When sent within a NEW message, this IE is used to ask a remote peer to automatically answer a call	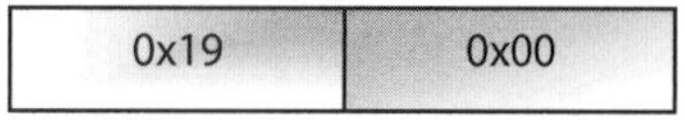 Figure 5.26 AUTOANSWER IE
Music on Hold MUSICONHOLD	This IE is used to request playing of music-on-hold while a call is in the QUELCH state. For more details about QUELCH message, refer to Chapter 6	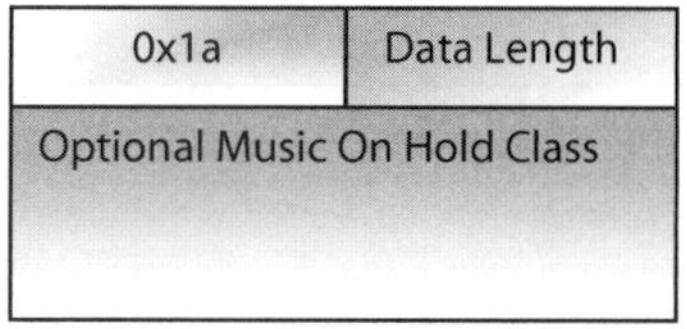 Figure 5.27 MUSICONHOLD IE
Transfer Identifier TRANSFERID	This IE identifies a transfer across all parties participating in a transfer operation. This number must be unique for the duration of the transfer process	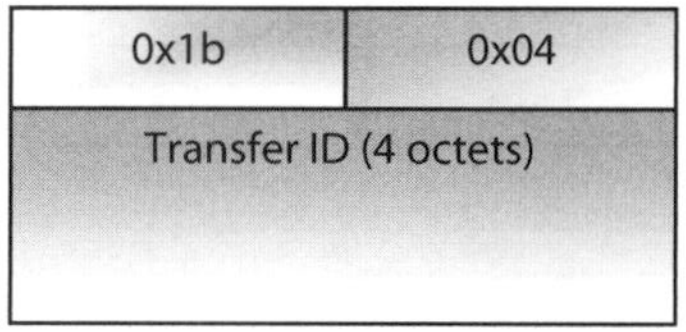 Figure 5.28 TRANSFERID IE

Referring DNIS
RDNIS

This IE is used to indicate the referring DNIS

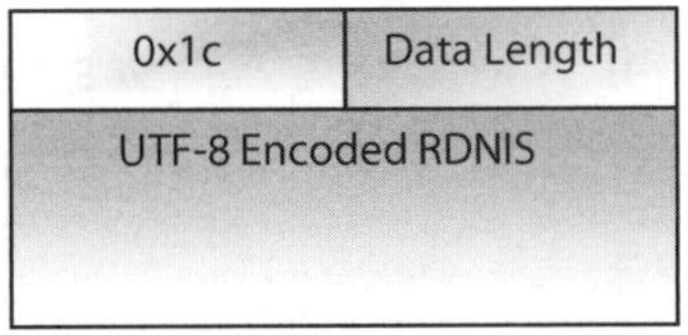

0x1c	Data Length
UTF-8 Encoded RDNIS	

Figure 5.29 RDNIS IE

Provisioning
PROVISIONING

For provisioning purposes, this IE is used to carry provisioning data destined to a given device

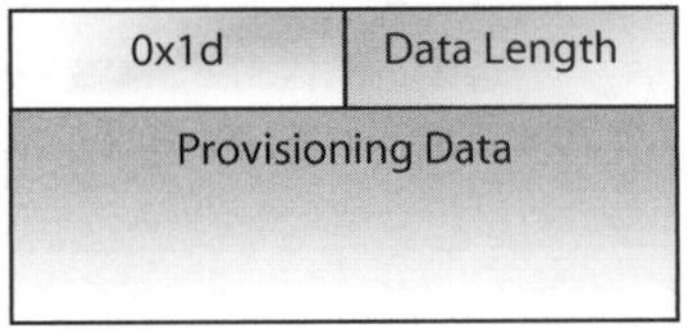

0x1d	Data Length
Provisioning Data	

Figure 5.30 PROVISIONING IE

AES Provisioning
AESPROVISIONING

This IE is used to convey AES encrypted provisioning data [AES]

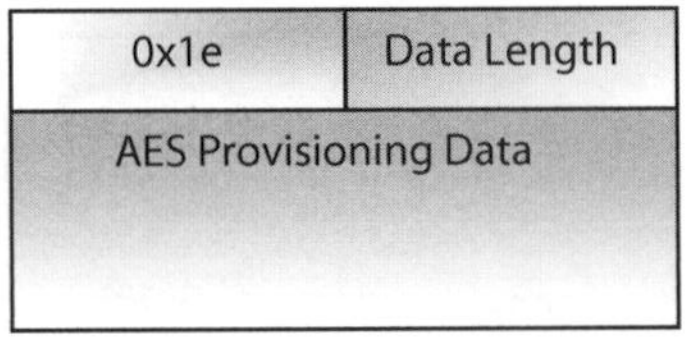

0x1e	Data Length
AES Provisioning Data	

Figure 5.31 AESPROVISIONING IE

Date-Time
DATETIME

In order to indicate the time when a given message is sent, an IAX peer may enclose the DATETIME IE in its IAX control messages, mainly in a NEW or REGACK message. More information about the values assigned to this IE field may be found at [IAX]

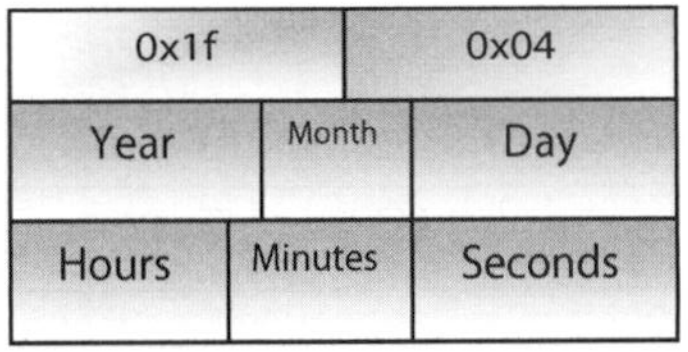

0x1f	0x04	
Year	Month	Day
Hours	Minutes	Seconds

Figure 5.32 DATETIME IE

(continued)

Table 5.1 (*Continued*)

Abbr. IE Name	Description	IE Message Format
Device Type DEVICETYPE	This IE is used by an IAX speaker to indicate the type of device requesting registration or firmware download	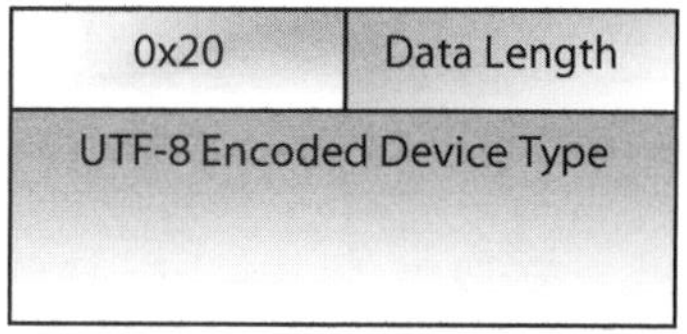 0x20 | Data Length UTF-8 Encoded Device Type **Figure 5.33** DEVICE TYPE IE
Service Identifier SERVICEIDENT	This IE is used to carry identifiers (to uniquely identify services) about a device requesting provisioning, so that appropriate data may be made available	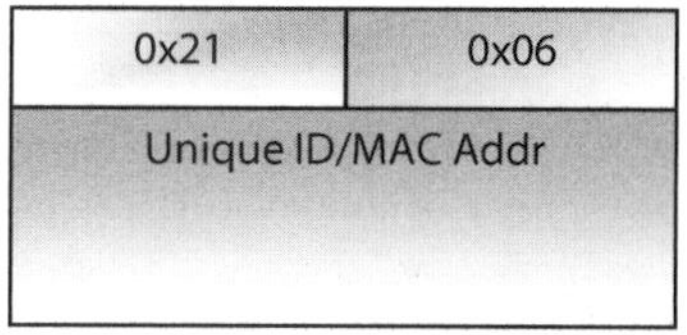 0x21 | 0x06 Unique ID/MAC Addr **Figure 5.34** SERVICE ID IE
Firmware Version FIRMWAREVER	This IE is used to indicate the version of the available firmware for a given device (depending on its type)	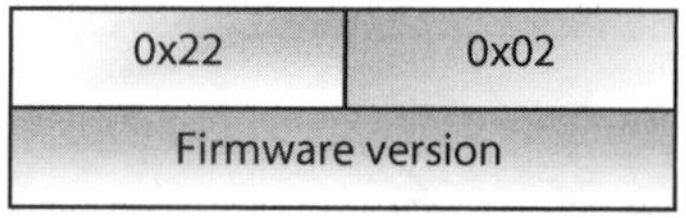 0x22 | 0x02 Firmware version **Figure 5.35** FIRMVER IE
Firmware Block Description FWBLOCKDESC	This IE is used to identify a block of firmware images. At least two cases may be considered: • If the IE is enclosed in an FWDOWNL message: this should be understood by the server as a request for the specified block of firmware • If the IE is enclosed in an FWDATA message: this should be understood as identification of the block of firmware carried in an FWBLOCKDATA IE	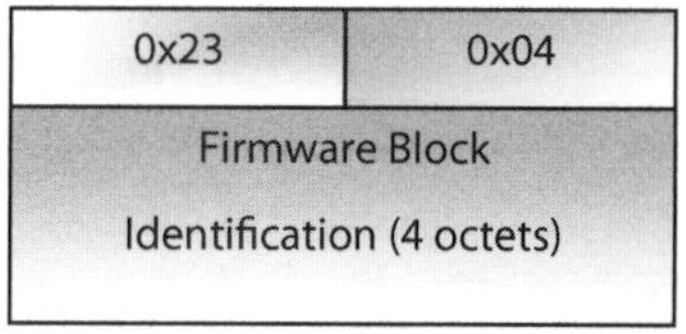 0x23 | 0x04 Firmware Block Identification (4 octets) **Figure 5.36** FW BLOCK DESC IE

Firmware Block Data
FWBLOCKDATA

This IE is used by a firmware server to deliver a block of firmware images to a given device

0x24	Data Length
Binary Block of Firmware Data	

Figure 5.37 FWBLOCK DATA IE

Provisioning Version
PROVVER

This IE is used to indicate the provisioning version of the software used

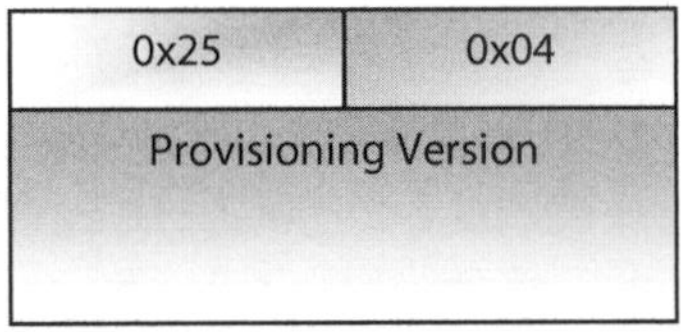

0x25	0x04
Provisioning Version	

Figure 5.38 PROVVER IE

Calling Presentation
CALLINGPRES

This IE indicates the calling presentation of a caller participant. The following values may be assigned to the Calling Presentation filed:

- 0x00: Allowed user/Number not screened
- 0x01: Allowed user/Number passed screen
- 0x02: Allowed user/Number failed screen
- 0x03: Allowed network number
- 0x20: Prohibited user/Number not screened
- 0x21: Prohibited user/Number passed screen
- 0x22: Prohibited user/Number failed screen
- 0x23: Prohibited network number
- 0x43: Number not available

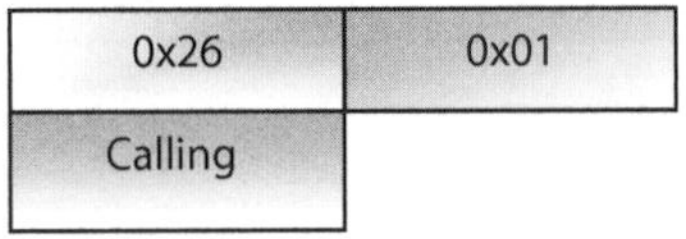

0x26	0x01
Calling	

Figure 5.39 CALLING PRES IE

(*continued*)

Table 5.1 (*Continued*)

Abbr. IE Name	Description	IE Message Format
Calling Type of Number CALLINGTON	This IE is used to indicate the Calling Type of Number (caller side), according to ITU-T Recommendation Q.931 specifications. Accepted Calling TON values are: • 0x00: Unknown • 0x10: International Number • 0x20: National Number • 0x30: Network Specific Number • 0x40: Subscriber Number • 0x60: Abbreviated Number • 0x70: Reserved for extension	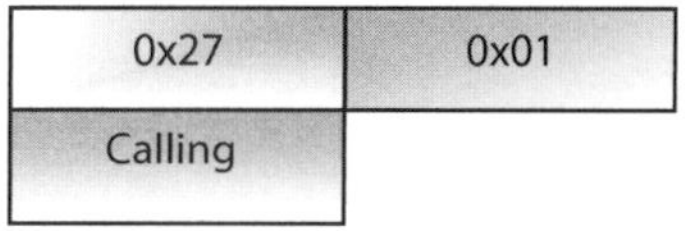 **Figure 5.40** CALLING TON IE

Calling Transit Network
CALLINGTNS
	This IE indicates the calling transit network selected to set up a call according to ITU-T Recommendation Q.931 specifications. Accepted CALLINGTNS IE values are:
	
	• TON: this field stores the Type of Network, as indicated below:
	  • 000: User-specified
	  • 010: National Network Identification
	  • 011: International Network Identification
	• Plan: this field stores the network identification plan according to the following values:
	  • 0001: Caller Identification Code
	  • 0000: Unknown
	  • 0011: Data Network Identification Code
	• Net ID: this field stores the actual network identification
		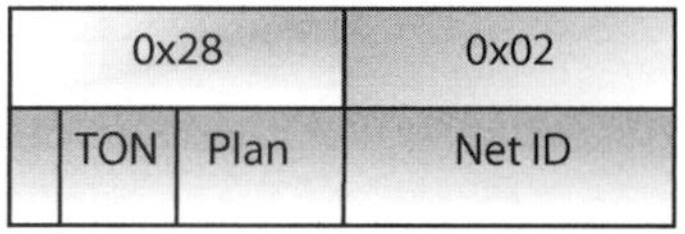
		
		Figure 5.41 CALLING TNS IE

Sampling Rate
SAMPLINGRATE

This IE is used to indicate to a remote peer the sampling rate to be used to send audio data. If this IE is not specified or not supported, the default sampling rate is 8 kHz

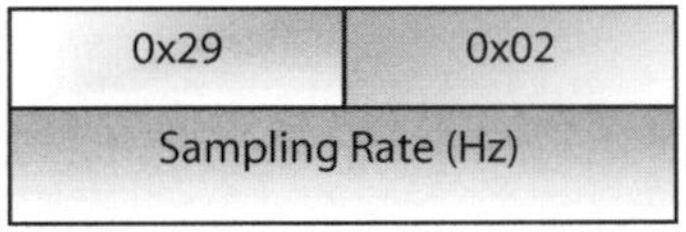

0x29	0x02
Sampling Rate (Hz)	

Figure 5.42   SAMPLING RATE IE

Cause Code
CAUSECODE

This IE is used to indicate the reason for rejection/hangup of a call. Cause code examples include (a full list can be found at [IAX]):

- 2: No route to specified transit network
- 3: No route to destination
- 41: Temporarily failure
- 42: Switch congestion

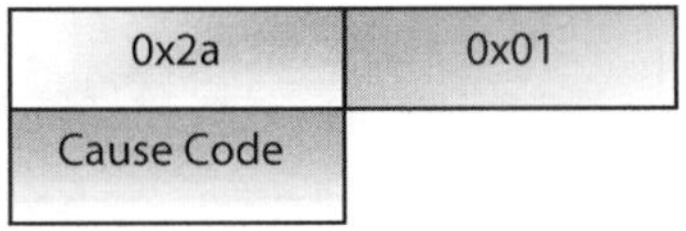

0x2a	0x01
Cause Code	

Figure 5.43   CAUSE CODE IE

Encryption
ENCRYPTION

This IE indicates the supported encryption methods of a given peer. The current accepted value is:

- 0x0001: AES-128

0x2b	0x02
Encryption Methods	

Figure 5.44   ENCRYPTION IE

(continued)

Table 5.1 (Continued)

Abbr. IE Name	Description	IE Message Format
Encryption Key ENCKEY	This IE is used to share an encryption key with a remote peer. This IE may be sent only in encrypted data	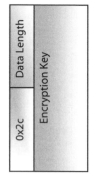 **Figure 5.45** ENCKEY IE
CODEC Preference CODEC PREFS	The objective of this IE is to indicate the CODEC preference to a remote peer. Therefore, an ordered list of CODECs may be carried in this IE. When this IE is not supported, the CAPABILITY or FORMAT IEs may be used to negotiate the CODEC to be used for a given call	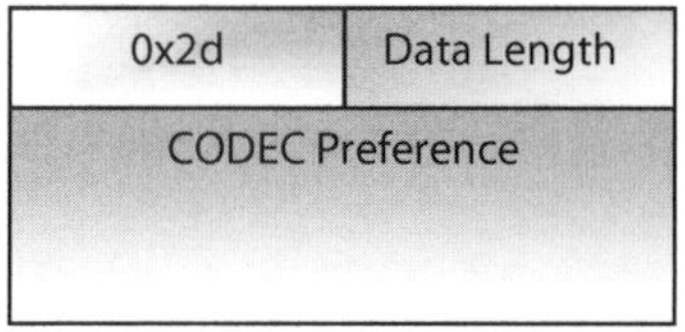 **Figure 5.46** CODEC PREFS IE
Jitter RR JITTER	This IE indicates the received jitter (mainly the current measured jitter) for a given call	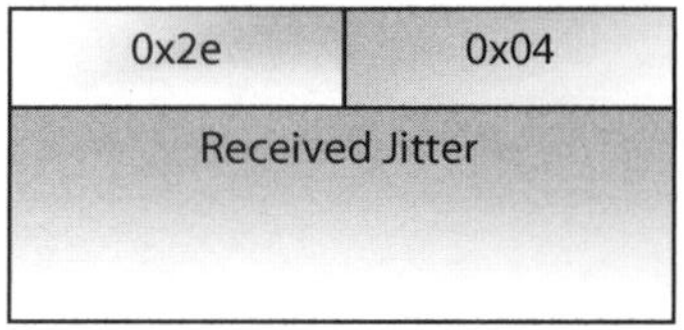 **Figure 5.47** RR JITTER IE

Figure 5.45 ENCKEY IE — 0x2c | Data Length / Encryption Key

Figure 5.46 CODEC PREFS IE — 0x2d | Data Length / CODEC Preference

Figure 5.47 RR JITTER IE — 0x2e | 0x04 / Received Jitter

Loss
RR LOSS

This IE carries the loss-related information for a given call. It includes the percentage and the count of lost frames

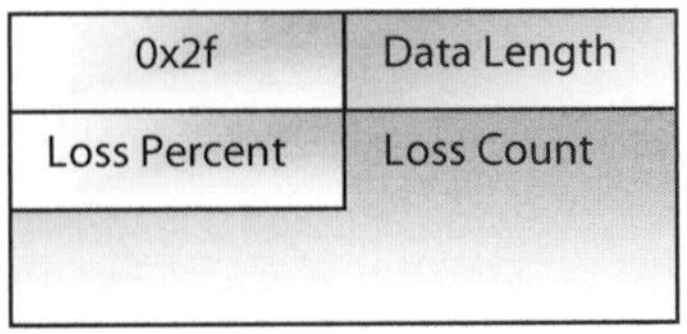

0x2f	Data Length
Loss Percent	Loss Count

Figure 5.48   RR LOSS IE

Received Frames
RR PKTS

This IE carries the total number of received frames for a given call

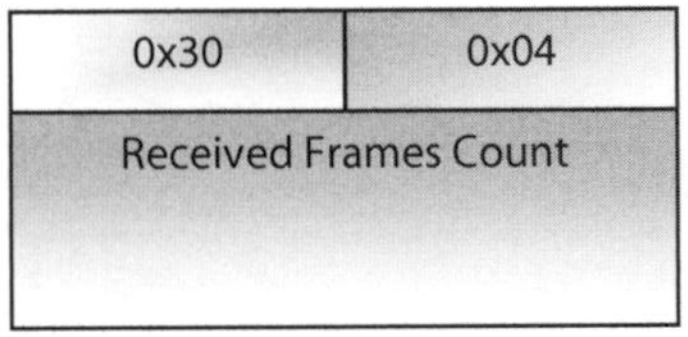

0x30	0x04
Received Frames Count	

Figure 5.49   RR PKTS IE

Delay
RR DELAY

This IE carries the number of milliseconds a frame may be delayed before it must be discarded. This is also called the timeout delay

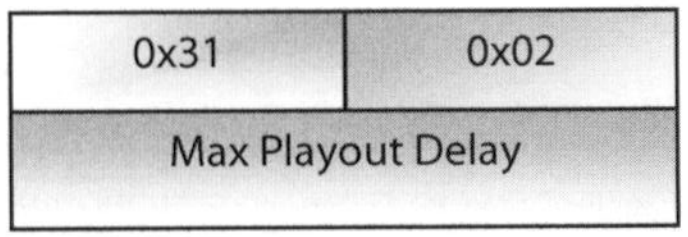

0x31	0x02
Max Playout Delay	

Figure 5.50   RR DELAY IE

Dropped Frames
RR DROPPED

The objective of this IE is to indicate the number of dropped frames within a given call

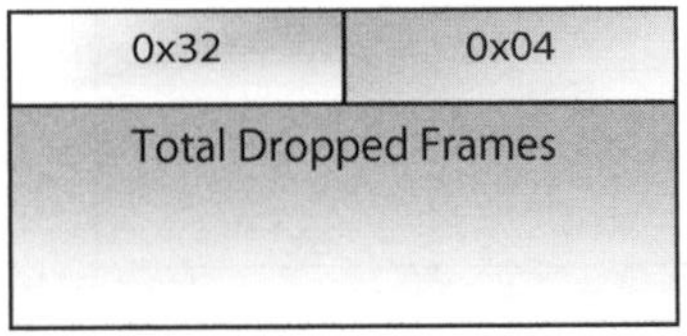

0x32	0x04
Total Dropped Frames	

Figure 5.51   RR DROPPED IE

(continued)

Table 5.1 (*Continued*)

Abbr. IE Name	Description	IE Message Format
Received Out of Order Frames RR OOO	The objective of this IE is to indicate the number of frames received out of order for a call	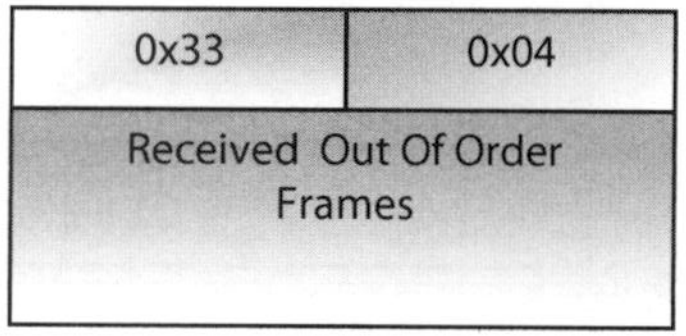

Figure 5.52 RR OOO IE

---

[1] For more information about ANI, please refer to [ANI].

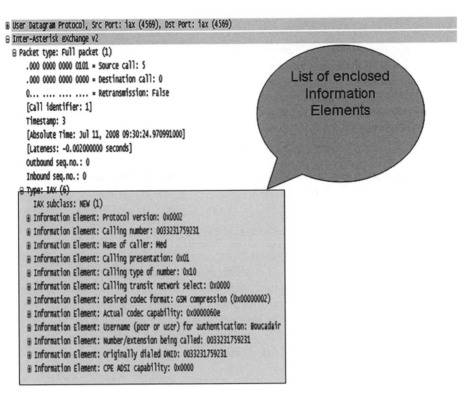

Figure 5.53 Example of an IAX message enclosing information elements.

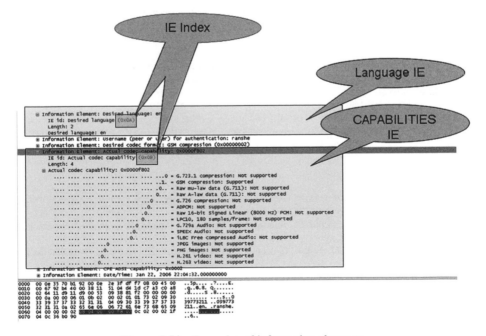

Figure 5.54 Examples of information elements.

5.2 List of IAX Information Elements

Table 5.1 groups the information elements defined in IAX specifications. The following data is provided for each information element:

- *IE Name/Abbreviated Name:* this column indicates both the full and the abbreviated name of the IE, as defined in [IAX].
- *Description:* provides a brief description of the IE and its possible uses.
- *IE message Format:* illustrates the format/structure of the IE.

For some IEs, a reference is made to the IAX message which carries that IE. For more details about these IAX messages, refer to Chapter 6.

5.3 Example of IAX Information Element Traces

Figure 5.53 illustrates an IAX message enclosing a set of information elements. Examples of enclosed IEs are: **Protocol Version IE, Calling Number IE, CODEC IE** and so on.

Figure 5.54 focuses on two information elements enclosed in IAX messages. The first is the **LANGUAGE IE**. As shown in this figure, the index of this IE is **0x0a** and its length is **2**. This IE indicates that the supported language is **English**. The second IE is the **CAPABILITIES IE**. This IE indicates the set of CODECs supported by the issuing IAX peers. The index of this IE is **0x08**. The issuing IAX peer indicates that only the GSM CODEC is supported and usable for forthcoming conversations.

References

[AES] US Department of Commerce/NIST, 'FIPS-197, Announcing the Advanced Encryption Standard', November 2001.

[ANI] Alliance for Telecommunications Industry Solutions, 'Automatic Number Identification (ANI) Digit Codes', September 1998.

[E164] ITU-T, 'The International Public Telecommunication Number Plan', Recommendation E.164, May 1997.

[IAX] Spencer, M., Shumard, K., Capouch, B. and Guy, E.,'IAX2: Inter-Asterisk eXchange Version 2', draft-guy-iax-04, work in progress.

[MD5] Rivest, R.,'The MD5 Message-Digest Algorithm', RFC 1321, April 1992.

[RSA] Kaliski, B. and Staddon, J.,'PKCS #1: RSA Cryptography Specifications Version 2.0', RFC 2437, October 1998.

[UTF8] Yergeau, F.,'UTF-8, a transformation format of ISO 10646', STD 63, RFC 3629, November 2003.

Further Reading

ITU-T Q.931 Recommendation, 'Q.931: ISDN User–Network Interface Layer 3 Specification for Basic Call Control', available at: http://www.itu.int/rec/T-REC-Q.931-199805-I/en.

Rosenberg, J., Schulzrinne, H., Camarillo, G., Johnston, A., Peterson, J., Sparks, R. et al., 'SIP: Session Initiation Protocol', RFC 3261, June 2002.

Rosenberg, J., Schulzrinne, H. and Kyzivat, P.,'Indicating User Agent Capabilities in the Session Initiation Protocol (SIP)', RFC 3840, August 2004.

6

IAX Messages

6.1 Introduction

Chapters 4 and 5 introduced the format of IAX messages and enclosed information element objects from a structure perspective. This chapter defines IAX messages from a functional standpoint. Therefore, a detailed description of valid IAX messages and the context of their invocation is provided, following the introduction of a tentative taxonomy. Several classification options may be used to group IAX control messages.

This chapter is structured as follows:

- Section 6.2 introduces a set of valid classification options for IAX control messages.
- Section 6.3 focuses on a functional classification of IAX messages. Thus, IAX control messages are divided into two groups: IAX requests and IAX responses.
- Section 6.4 classifies IAX control messages into several functional groups with regards to their invocation context.
- Section 6.5 is dedicated to IAX media messages.
- Section 6.6 groups IAX control messages into two categories: reliable and unreliable.

6.2 Taxonomy of IAX Messages

IAX protocol defines a set of messages which may be invoked for numerous uses, such as initiating, terminating or controlling call sessions, sending media streams and so on. These messages may be classified into several categories, depending on the adopted criteria.

Within this section, four criteria for the classification of IAX messages are adopted and elaborated. These criteria are:

- *IAX Requests vs IAX Responses*: the logic behind this criterion is to classify IAX messages according to their function with regards to client/server logic; that is, to determine whether a given IAX message is a 'request' or a 'response' to an already issued IAX 'request'. Note that this criterion does not cover media messages but only signalling IAX messages (also know as IAX control messages; see Section 6.3).

Inter-Asterisk Exchange (IAX): Deployment Scenarios in SIP-Enabled Networks   Mohamed Boucadair
© 2009 John Wiley & Sons, Ltd

- *Functional Objectives*: all IAX messages do not serve the same functional objective(s). Section 6.3.3 describes the types of functional objective and the messages that belong to each one.
- *Signalling vs Media IAX Messages*: as indicated above, IAX is an all-in-one protocol. IAX can be used for signalling purposes and also to convey media streams. Section 6.5 focuses on media messages, since Section 6.3 focuses only on signalling messages.
- *Reliable vs Unreliable*: IAX messages may be reliable or unreliable (see Section 6.6). Reliable messages require an acknowledgement from the receiving IAX peer.

6.3 IAX Requests/Responses

This section focuses on the first classification option and discusses IAX messages from a client/ server logic.

6.3.1 IAX Requests

Table 6.1 lists the IAX request messages. It also provides the following information:

- *IAX Message/Message Full Name*: the abbreviated and full name of an IAX request message.
- *Description*: a brief description of the objectives and the invocation context of a given IAX request.
- *Enclosed IE*: the information elements which may be sent in an IAX request message. Inclusion of these inforation elements may be mandatory, optional or conditional.

Table 6.1   List of IAX request messages

IAX Message *Message Full Name*	Description	Enclosed IEs
AUTHREQ *Authentication Request*	This message is sent as a response to a NEW message if authentication is required for the call to be accepted. When receiving an AUTHREQ, the remote peer must respond with an AUTHREP or HANGUP message	USERNAME, CHALLENGE and/or AUTHMETHODS
DIAL	This message may be used by an IAX peer which does not maintain its dial plan. Once extension is determined by the exchange of DPREQ and DPREP, it is enclosed in a DIAL message	CALLEDNUMBER and/or CALLEDCONTEXT
DPREQ *Dial Plan Request*	In order to determine the number to call according to a dial plan maintained by a server, an IAX speaker may issue a DPREQ	CALLEDNUMBER
FLASH	This message's purpose is to notify the remote IAX peer that a mid-call event has occurred in the local terminal. The interpretation of this event is system-dependent. For instance, this message may be generated when an analogue telephone	No IE

(continued)

Table 6.1 (*Continued*)

IAX Message *Message Full Name*	Description	Enclosed IEs
	adapter has made a circuit interruption during an answered call	
FWDOWNL *Firmware Download*	In order to request a firmware download, an IAX device has to issue a FWDOWNL message to the firmware server. Indications such as device type, firmware version and so on may be carried in the FWDOWNL message. The firmware server has to use the received IE to determine the firmware to be transferred to the requesting device	DEVTYPE and FWBLOCKDATA
HANGUP	In order to tear down an already established call, an IAX speaker has to send a HANGUP message to the remote participating IAX peer. This message may indicate the cause of the teardown, if any. The call context in the remote IAX peer has to be destroyed and acknowledged with an ACK message	CAUSE and/or CAUSECODE
HOLD	The purpose of this message is to request the remote peer to stop sending media streams. If this feature is not supported by the remote peer, this request should be ignored. From an IAX protocol perspective, this message may be accepted only for calls initiated by a DIAL request	No IE
LAGRQ *Lag Request*	In order to evaluate the lag between two peers, a LAGRQ may be sent to a remote IAX speaker. The lag between the two peers is computed by comparing the timestamp of the LAGRQ and the time the LAGRP was received	No IE
MWI *Message Waiting Indicator*	This message is used to indicate to a given user that one or more messages are waiting. This message may enclose the amount of new and/or old messages	MSGCOUNT
NEW	In order to initiate a call, a NEW message has to be issued to the remote peer or to a third participant. NEW messages are differentiated in a given IAX peer by the 'Source Call Number', which must be enclosed in the NEW message, and which must be maintained during the call establishment exchange. Note that the Destination Call Number has to be set to 0. Several information elements may be inserted in a NEW message, for example VERSION, CAPABILITY and CALLINGNUMBER. The NEW message may indicate a single CODEC or a list of	VERSION, CALLEDNUMBER, AUTOANSWER, CODECPREF, CALLINGPRES, CALLINGNUMBER, CALLINGTON, CALLINGTNS, CALLINGNAME, ANI, LANGUAGE, DNID, CALLEDCONTEXT, USERNAME, RSARESULT, MD5RESULT, FORMAT, CAPABILITY, ADSICPE and/or DATETIME

(*continued*)

Table 6.1   (*Continued*)

IAX Message Message Full Name	Description	Enclosed IEs
	CODECs for the media exchange. This message may be answered with a REJECT, AUTHREQ, HUNGUP or ACCEPT message	
PING	The main purpose of this message is to test the connectivity between two IAX peers. The exchange of PING message may be conditioned by a timer value (which by default is set to 20 seconds). Upon the receipt of a PING message, the remote peer should reply with a PONG message	No IE
POKE	This message is similar to the PING request. Nevertheless, it may be sent only when there is no existing call to the remote IAX peer	No IE
PROVISION	In order to deliver provisioning data, a PROVISION message may be sent to an IAX speaker. This enclosed provisioning data format is device-specific and is out of the scope of IAX specification	PROVISIONING and/or AESPROVISIONING
QUELCH	This message is similar to a HOLD message but should be used only for calls which have been initiated by a NEW message. Therefore, this message may be sent only after the exchange of an ACCEPT message	No IE
REGREL *Registration Release*	In order to delete a registration record maintained by an IAX registrar, an IAX registrant has to issue a REGREL message including the user name and optionally the cause of the registration release. The IAX registrar has to answer the registrant with a REGAUTH if authentication is required, or with a REGACK to acknowledge the registration record has been deleted	MD5RESULT, RSARESULT, CAUSE and/or CAUSECODE
REGREQ *Registration Request*	In order to be reachable from other remote IAX peers, an IAX speaker should register itself within a registrar server by using a REGREQ message. This message specifies the user name and, optionally, an expire timeout. When authentication is required by the registrar, a REGAUTH reply has to be sent back to the IAX registrant. The registrar may also reject the registration request by issuing a REGREJ message, or acknowledge that the registration has been proceeded successfully by sending a REGACK message	USERNAME, MD5RESULT, RSARESULT and/ or REFRESH

(*continued*)

Table 6.1 (*Continued*)

IAX Message *Message Full Name*	Description	Enclosed IEs
TRANSFER	In order to be reachable with another call number, an IAX peer may specify the new number. The call establishment has to restart using the new call number. A HUNGUP message should then be issued after a TRANSFER message in order to destroy the previous call context	CALLEDNUMBER and/or CALLEDCONTEXT
TXREQ *Transfer Request*	In order to request a transfer operation, an IAX peer has to issue a TXREQ message for both IAX peers involved in the call. The transfer operation is uniquely identified by a transfer identifier which should be maintained during the transfer operation. A TXREQ message also includes the IP address of the remote peer to which the call should be transferred	APPARENTADDR, CALL-NUMBER and/or TRANDERID
UNHOLD	When media transmission has been stopped between two IAX peers after an exchange of a HOLD message, the two peers may restart sending media by generating an UNHOLD message	No IE
UNQUELCH	This message is similar to an UNHOLD message. It should be sent only if a HOLD message has been exchanged between two IAX peers	No IE
VNAK *Voice Negative Acknowledgement*	When receiving an OOO (Out of Order) message, the local IAX peer should issue a VNAK message. Upon receipt of the VNAK message, the remote peer must retransmit all messages up to the one identified by the sequence number indicated in the VNAK message	No IE

6.3.2 IAX Responses

Table 6.2 lists the IAX response messages. It also provides the following information:

- *IAX Message/Message Full Name*: the abbreviated and full name of an IAX response message.
- *Description*: a brief description of the objectives and the uses of a given IAX response.
- *Enclosed IE*: the information elements which may be sent in an IAX response message. Inclusion of these information elements may be mandatory, optional or conditional.

Table 6.2 List of IAX response messages

IAX Message Message Full Name	Description	Enclosed IEs
ACCEPT	This message is issued by an IAX peer to accept a call establishment request. It indicates the CODEC selected for the media exchange. The selected CODEC must be supported by the remote peer	FORMAT
ACK *Acknowledgement*	This message is sent to a remote IAX speaker to acknowledge the receipt of a full frame. An ACK message must particularly be sent when a NEW, HANGUP, REJECT, ACCEPT, PONG, AUTHREP, REGREL, REGACK, REGREJ or TXREL message has been received by the local IAX speaker	No IE
ANSWER	This message is sent by an IAX peer in order to indicate that it has accepted the call-establishment request and will start to send media to the remote participant. This IAX peer will open media channels and stop sending any mid-call messages	No IE
AUTHREP *Authentication Reply*	This message is issued by an IAX speaker in answer to an AUTHREQ request. It should enclose appropriate challenge-response or password information. Upon receipt of the AUTHREP message by the remote peer, the farmer must answer with an ACCEPT or a REJECT message	RSARESULT or MD5RESULT
DPREP *Dial Plan*	This message is a response to a DPREQ. It notifies the remote peer about the status of the requested number in a remote dial plan routing	CALLEDNUMBER, DPSTATUS, DP REFRESH
FWDATA *Firmware Data*	After receiving a firmware download request, the firmware server may respond with several FWDATA messages in order to convey firmware binary blocks. To notify the end of the firmware download, the last FWDATA has to enclose a data block with length 0	FWBLOCKDESC and FWBLOCKDATA
INVAL	This message is sent when an IAX message has been received after the call context has been destroyed, for instance when a message is received after a HANGUP message has been issued by an IAX peer. When the remote peer receives this message, it must destroy the call context at its side	No IE
LAGRP *Lag Response*	Upon receipt of a LAGRQ, an IAX speaker should reply with a LAGRP message, after passing this message through the jitter buffer of the remote IAX peer. The LAGRP has the same timestamp as the LAGRQ	No IE
PONG	Upon the receipt of a PING or a POKE message, an IAX peer should response with a PONG message	RRJITTER, RRPKTS, RRDELAY and/ or RRDROPPED
PROCEEDING	During the processing of a call request, an intermediary IAX node may use a PROCEEDING	No IE

(continued)

Table 6.2 (*Continued*)

IAX Message Message Full Name	Description	Enclosed IEs
	message to notify the requesting participant that the request has been forwarded to a third participant and that an answer has not yet been received	
REGACK *Registration Acknowledgment*	This message is sent to acknowledge the receipt of a REGREQ specifying the expire timer of the registration record. When no timeout has been specified, the default value must be set to 60 seconds. Additional information may be carried in a REAGACK message as indicated in the Enclosed IEs column	USERNAME, DATE-TIME, APPAREN-TADDR, MSGCOUNT, CALL-INGNUMBER CALL-INGNAME, FIRM-WAREVER and/or REFRESH
REGAUTH *Registration Authentication*	When authentication is required for a registration procedure, a REGAUTH message is sent back upon receipt of a REGREQ or a REGREL message. Authentication methods and challenges are conveyed in the body of the REGAUTH to indicate to the remote IAX speaker the authentication method to be used and that it should be challenged. The remote IAX speaker has to answer with a new REGREQ or REGREL message including the result of the authentication method applied to the received challenge	USERNAME, AUTH-METHODS and/or CHALLENGE
REGREJ *Registration Rejection*	In order to reject a registration request or release request, an IAX registrar server has to send back a REGREJ message. This message may enclose the cause of the registration reject so as to notify the registrant IAX speaker	CAUSE and CAUSECODE
REJECT	If a remote IAX peer has to decline a NEW, AUTHREP, DIAL or ACCEPT message, it has to send a REJECT message. The cause of the reject event may be indicated in the body part of the REJECT message. Examples of the reason for rejection may be authentication, CODEC mismatch and so on. An IAX speaker who has received a REJECT message has to destroy the call context and replay with an ACK message	CAUSE and/or CAUSECODE
RINGING	A remote IAX peer has to send a RINGING message to indicate to the remote peer that the IAX call establishment request has been received and that it is in process of validating/rejecting the call setup request	No IE
TXACC *Transfer Accept*	This message is a response to a TXCNT request indicating that it should be accepted. A transfer identifier is enclosed, which must be maintained during the transfer process. The remote peer which receives this message should stop sending media to the current location and send a TXREADY message in response	TRANSFERID

(*continued*)

Table 6.2 (*Continued*)

IAX Message *Message Full Name*	Description	Enclosed IEs
TXCNT *Transfer Connectivity*	This message is used to verify connectivity with the new peer to which a call should be transferred. In order to identify the transfer operation, an identifier is enclosed in this message. This identifier must be maintained during the transfer process. The remote peer which receives this message should send a TXACC message in response	TRANSFERID
TXREADY *Transfer Ready*	After verification of the connectivity to which a call should be transferred, a local IAX peer has to issue a TXREADY message. This message must enclose the transfer identifier so as to ensure that this peer is part of the transfer operation. In order for the transfer operation to be successful, both intervening parties should answer with a TXREADY message	TRANSFERID
TXREJ *Transfer Rejection*	During a transfer operation, this message may be sent to indicate that a given peer cannot execute this operation. Other TXREJ messages may be generated so as to inform the remote peer that the transfer operation has been rejected	No IE
TXREL *Transfer Release*	In order to indicate that a transfer operation has been successfully handled, a TXREL message should be issued. This message should be acknowledged by an ACK message	CALLEDNUMBER
UNSUPPORT	Upon receipt of an unsupported IAX message, an UNSUPPORT message should be sent to inform the remote peer that it is not supported	UNKNOWN

6.3.3 Information Elements and IAX Messages

As already mentioned, one or several information elements may be enclosed in a given IAX message. Table 6.3 describes for each IAX IE the IAX messages that may carry it.

The structure of Table 6.3 is as follows:

- *IE Name*: indicates both the full and abbreviated name of the corresponding information element, as defined in Chapter 5.
- *Related IAX Messages*: lists the IAX messages which may/must enclose this information element. Enclosing a given information element can be 'mandatory', 'conditional' or 'optional'. Within this book, this level of granularity is not provided. Protocol implementers are invited to refer to [IAX] for more details.

Table 6.3 Candidate IEs to be conveyed in IAX messages

IE Name	Related IAX Messages
CALLED NUMBER	NEW, DPREQ, DPREP, DIAL and TRANSFER
CALLING NUMBER	NEW
CALLING ANI	NEW
CALLING NAME	NEW
CALLED CONTEXT	NEW or TRANSFER
USERNAME	NEW, AUTHREQ, REGREQ, REGAUTH or REGACK
CAPABILITY	NEW
FORMAT	NEW or ACCEPT
LANGUAGE	NEW
VERSION	NEW
ADSICPE	NEW
DNID	NEW
AUTHMETHODS	AUTHREQ or REGAUTH
CHALLENGE	AUTHREQ or REGAUTH
MD5RESULT	AUTHREP or REGREQ
RSARESULT	AUTHREP or REGREQ
APPARENT ADDR	REGACK or TXREQ
REFRESH	REGREQ, REGACK and DPREP
DPSTATUS	DPREP
CALLNO	TXREQ, TXREADY or TXREL
CAUSE	HANGUP, REJECT, REGREJ and TXREJ
IAX2 UNKNOWN	Unsupported IAX2 messages
MSGCOUNT	REGACK
AUTOANSWER	NEW
MUSICONHOLD	QUELCH
TRANSFERID	TXREQ or TXCNT
PROVISIONING	PROVISION
AESPROVISIONING	PROVISION
DATETIME	NEW and REGACK
DEVICETYPE	FWDOWNL or REGREQ
SERVICEIDENT	REGREQ
FIRMWAREVER	REGACK
FWBLOCKDESC	FWDOWNL and FWDATA
FWBLOCKDATA	FWDATA
PROVVER	REGREQ
CALLINGPRES	NEW
CALLINGTON	NEW
CALLINGTNS	NEW
SAMPLINGRATE	NEW or ACCEPT
CAUSECODE	HANGUP, REJECT, REGREJ or TXREJ
ENCRYPTION	NEW or AUTHREQ
ENCKEY	Any IAX2 message invoked for authentication
CODEC PREFS	NEW
RR JITTER	PONG
RR LOSS	PONG
RR PKTS	PONG
RR DELAY	PONG
RR DROPPED	PONG
RR OOO	PONG

6.4 IAX Functional Categories

IAX messages may also be classified according to their objectives and accomplished functions, as detailed in Table 6.4.

This table is structured as follows:

- *Message Category*: provides the name of each IAX message category.
- *Status:* IAX functions may be 'optional' or 'mandatory'. This column describes the status of each category.
- *Description*: provides a description of the function accomplished by each category.
- *Members*: lists the messages which belong to each category.

Note that not all signalling IAX messages belong to one of these functional categories. More precisely, **ACK**, **INVAL**, **VNAK**, **MWI** and **UNSUPPORT** cannot be classified according to the functional criteria.

Table 6.4   IAX message categories

Message Category	Status	Description	Members
Registration	Optional	The role of the members of this category is to manage registration operations, mainly requesting, validating, releasing and acknowledging location information	REGREQ, REGAUTH, REGACK, REGREJ and REJREL
Call Leg Management	Mandatory	The main objective of this category is to initiate, accept and terminate a call session	NEW, ACCEPT, REJECT, HANGUP, AUTHREP and AUTHREQ
Call Control	Mandatory	These messages aim to provide the status of an ongoing call-session request	PROCEEDING, RINGING and ANSWER
Mid-Call Link Operations	Optional	These messages may be issued when a call is activated	FLASH, HOLD, UNHOLD, QUELCH, UNQUELCH and TRANSFER
Call Path Optimisation	Optional	The purpose of these message is to ensure that decoupling media and signalling paths may be enforced without major pain	TXREQ, TXCNT, TXACC, TXREADY and TXREJ
Network Monitoring	Mandatory	The purpose of these messages is to maintain the active session and evaluate the network IP QoS parameters, such as delay, loss and jitter	POKE, PING, PONG, LAGRQ and LAGRP
Digit Dialing	Optional	This group of messages is issued when a local IAX speaker does not maintain a dial-plan local routing. These messages help to invoke another IAX device to retrieve the number to be dialed	DPREQ, DPREP and DIAL

(continued)

Table 6.4   (*Continued*)

Message Category	Status	Description	Members
Provisioning	Optional	This category's main goal is to provide provisioning data to a given IAX device	PROVISION
Firmware Download	Optional	Firmware download messages ease checking and downloading of new firmware versions compliant with device capabilities. The device capabilities may be enclosed as information elements	FWDOWNL and FWDATA

6.5 IAX Media Frames

Unlike SIP (Session Initiation Protocol, [SIP]), which relies on RTP (Real-Time Transport Protocol, [RTP]) to send media flows, IAX supports media-exchange features. Therefore, several media types are supported by the IAX protocol.

IAX media messages are usually exchanged as mini frames, even if the protocol recommends issuing a full frame when the timestamp reaches a multiple of **32 786**. The reason behind this recommendation is to enhance the reliability of the media exchanged between two IAX peers and to check that the remote IAX peer is still alive.

The use of mini frames as an envelope for exchanging media streams aims at optimising the available bandwidth (IAX mini frames are smaller than full frames).

As far as the transport layer is concerned, media IAX frames are exchanged using the same UDP (User Datagram Protocol, [UDP]) port number used to transmit signalling messages.

Table 6.5 lists the current media frames as defined in [IAX].

6.6 IAX Reliable/Unreliable Messages

IAX protocol defines two types of message:

- *Reliable Messages*: these acknowledgement from the remote IAX peer. All full frame messages are reliable messages. These reliable messages are retransmitted until they are

Table 6.5   IAX media messages

Frame Name	Description
DTMF	Used to carry a single digit of DTMF
Voice	Used to carry voice data (also indicates the used CODEC)
Video	Used to carry video data (also indicates the used CODEC)
Text	Used to carry a text IAX message
Image	Used to convey a single image
HTML	Used to convey HTML data
Comfort Noise	Used to convey the comfort noise to be played

acknowledged. To do so, a timer is maintained by each IAX speaker. After expiration of this timer, the IAX speaker has to retransmit the message until an acknowledgement message is delivered or the maximum retry limit is reached.

* *Unreliable messages*: these are not acknowledged upon reception. All mini frames and meta frames are unreliable IAX messages. These messages are used to convey media streams.

From what is stated above, we conclude that all IAX full frames are reliable and only mini frames, meta videos and trunk frames are unreliable. In particular, all IAX signalling messages are reliable, except the **ACK** message.

References

[IAX] Spencer, M., Shumard, K., Capouch, B. and Guy, E., 'IAX2: Inter-Asterisk eXchange Version 2', draft-guy-iax-04, work in progress.
[RTP] Schulzrinne, H., Casner, S., Frederick, R. and Jacobson, V., 'RTP: A Transport Protocol for Real-Time Applications', RFC 1889 (proposed standard), January 1996.
[SIP] Rosenberg, J., Schulzrinne, H., Camarillo, G., Johnston, A., Peterson, J., Sparks, R. et al., 'SIP: Session Initiation Protocol', RFC 3261, June 2002.
[UDP] Postel, J., 'User Datagram Protocol', RFC 768, August 1980.

Further Reading

Rosenberg, J. and Schulzrinne, H., 'Reliability of Provisional Responses in the Session Initiation Protocol (SIP)', RFC 3262, June 2002.

7

IAX Connectivity Considerations

7.1 Introduction

The following sections aims at describing how IAX connectivity is achieved. In particular, this chapter's objectives are to highlight:

- How IAX messages are transported.
- How IAX sessions are multiplexed using a single transport port.
- How reliability is ensured.
- How IAX messages are encrypted and authentication-enforced.
- How IAX handles NAT traversal.
- Why IAX is IP protocol-version agnostic.

A more detailed discussion of NAT traversal is given in Chapter 11.

7.2 IAX Transport Protocol

Most of the existing VoIP (Voice over IP) protocols use TCP (Transport Control Protocol, [TCP]) to convey signalling messages and UDP (User Datagram Protocol, [UDP]) to exchange media streams. The use of TCP for signalling messages is motivated by the need to ensure a high reliability of those messages and a robust transfer delivery. Indeed, if a signalling message is not received by a remote peer because it is lost in the crossed IP path, a retransmission procedure is undertaken until its effective delivery. As far as real-time media streams are concerned, this reliability requirement is no longer valid (i.e. retransmission of loosed messages is not required. This requirement is not to be confused with the need to have a non-'lossy' underlying IP network. Low loss [RFC2680] is required to offer VoIP with acceptable QoS (Quality of Service)). For these reasons, UDP is used to send media streams.

From an IAX perspective, all messages (including media and signalling) are sent over UDP. Reliability is ensured by an IAX proprietary mechanism (as described in Section 7.5).

Inter-Asterisk Exchange (IAX): Deployment Scenarios in SIP-Enabled Networks   Mohamed Boucadair
© 2009 John Wiley & Sons, Ltd

7.3 IAX Port Number

IAX uses a single port number to send and receive IAX messages from remote peers. IAX use the same port number for both signalling and media flows. The default port number value is **4569**. This port has been assigned by IANA (Internet Assigned Numbers Authority). In earlier stages, IAX used **5036** to place and receive IAX calls.

Unlike RTP (Real-Time Transport Protocol, [RTP]), no dynamic port assigned is enforced to send media flows. Only one single port is used to send and receive media traffic. Moreover, the control of the 'aliveness' of the remote peer is achieved in the same session, owing to invocation of dedicated IAX messages. Therefore, no additional port number (as required for RTCP (Real-Time Control Protocol, [RTP])) is required in the context of IAX deployment.

Figure 7.1 provides an overview of the structure of an IAX message, especially the transport protocol and both the source and the destination port numbers. In this figure, the default port number is used to issue the IAX message.

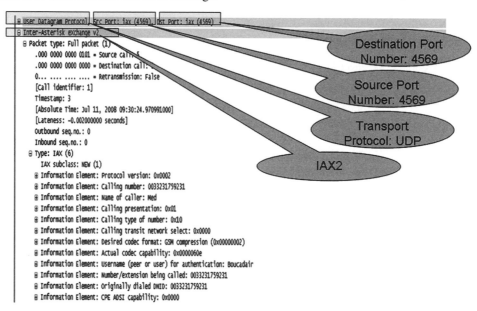

Figure 7.1   Example of IAX trace (transport protocol and port numbers)

This single-port approach to both signalling and media, combined with the minimal use of underplaying IP connectivity information at the IAX level (that is, IAX messages avoid carrying IP-related information such as an IP address or a port number), makes it easier to cross NAT (Network Adress Translator, [NAT]) and to configure firewall policies. Of course, NAT associations within a crossed NAT box must be refreshed by issuing regular IAX messages, so as to allow a successful IAX session with a given IAX participant behind a NAT box. The configuration of timeout associated with these NAT associations should be taken into account in the provisioning of IAX registration or refresh timers.

7.4 IAX Call Multiplexing and Demultiplexing

Figure 7.2 illustrates a scenario where four IAX-enabled user agents (UA) can communicate using IAX protocol.

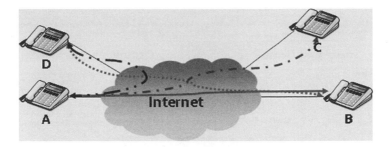

Figure 7.2 Multiplexing IAX sessions (1)

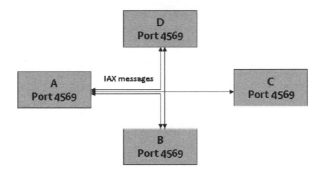

Figure 7.3 Multiplexing IAX sessions (2)

For illustration purposes, node **A** may maintain simultaneous IAX sessions with **D**, **C** and **B** using the same IAX port number.

These IAX sessions (that is, the ones maintained by **A** with remote IAX peers) use a single port number (**4569**) to send and receive signalling and media traffic for all active calls, as shown in Figure 7.3.

In this example (Figures 7.2 and 7.3), it is assumed that all involved IAX nodes use the IAX default port.

In order to distinguish data associated with each IAX session (that is, multiplexing and demultiplexing IAX sessions), an IAX-enabled user agent (UA) exploits the **Source Call Number** and **Destination Call Number** fields (indicated in Figure 7.4).

- **Source Call Number** is used to identify the session at the source side.
- Distinguishing between several active sessions with the same remote peer is possible through the **Destination Call Number** field.

Consequently, two IAX peers can initiate and maintain unambiguously several IAX sessions using the same UDP port number.

7.5 IAX Reliability Mechanism

Because IAX does not use TCP to convey its signalling messages, denoted also as 'control messages', reliability must be supported at the application layer so as to ensure the delivery of

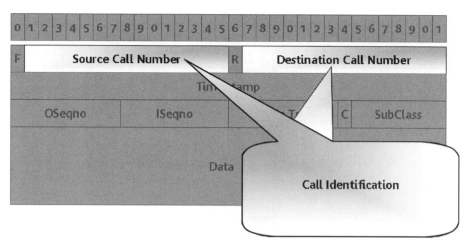

Figure 7.4   IAX field used to enforce multiplexing of IAX sessions

critical messages, which are used to initiate, terminate and manage multimedia sessions. This section provides a description of the reliability mechanisms as defined and supported by IAX implementations.

Indeed, IAX protocol distinguishes between two types of message:

- *Reliable Messages*: these require acknowledgement by the remote IAX peer. All full frame messages are reliable messages. These reliable messages are retransmitted until they are acknowledged. To do so, a timer is maintained by each IAX speaker. After expiration of this timer, the IAX speaker has to retransmit the message until it receives an acknowledgement message or until the maximum retry limit is reached.
- *Unreliable Messages*: these messages are not acknowledged upon receipt by a remote IAX peer. All mini frames and meta frames are unreliable. These messages are used to convey media streams.

Figure 7.5 highlights the useful fields for ensuring reliability.

For every ongoing call, each IAX participant maintains a **Timestamp**, Ongoing Sequence Number (**OSeqno**) and Ingoing Sequence Number (**ISeqno**).

- **Timestamp** is used to indicate the number of milliseconds since the initialisation of the call.
- **OSeqno** stores the number of messages sent to the remote IAX peer.
- **ISeqno** indicates the highest-number received message within a given call context.

All these counters are set to zero at the initialisation of an IAX call. **OSeqno** and **ISeqno** are incremented by **1** for all reliable messages, except for **ACK, INVAL, TXCNT, TXACC** and **VNAK,** which do not change these counters.

The order followed when checking received messages is based on the **Timestamp** value. Therefore, all frames which are out of order are ignored; a **VNACK** is sent to notify the sender participant of this. **ISeqno** is then used to acknowledge a reliable message.

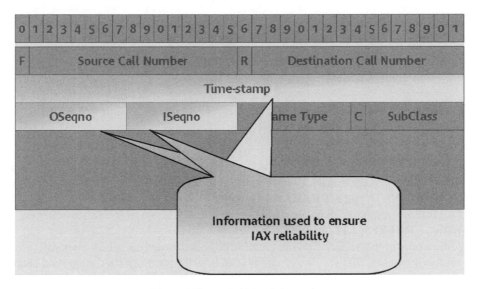

Figure 7.5   Reliability information

Figure 7.6 summarises the reliability-checking procedure as adopted by IAX implementations:

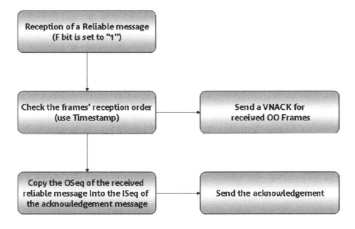

Figure 7.6   Reliability procedure

7.5.1 IAX Timers

An IAX-enabled user agent (also called IAX speaker, IAX peer or IAX endpoint) maintains several timers for its correct protocol operations, particularly:

- *Registration Timer*: used to condition issuing of registration refresh messages, so as to maintain a registration record within a registrar server.
- *Retransmission Timer*: used to condition the retransmission procedure when a given IAX message failed to be delivered.

- *Connectivity Timer*: used to issue a dedicated message to check whether a remote IAX peer is still alive and actually receives IAX messages.

These timers are described below.

7.5.2 Registration

As described in Chapter 6, IAX supports a registration procedure. This procedure aims to notify or inform a remote peer (usually a registrar server) about the availability of a local IAX peer. An explicit registration procedure is achieved using an explicit IAX message, mainly **REGREQ**. This message carries a set of useful information elements such as **USERNAME IE**, **CALLING NB IE** and a '**REFRESH IE**. Once received by the registrar, a response must be issued. This response must enclose an apparent-address information element and, optionally, a **REFRESH IE**.

In order to ensure that registration information provided by an IAX registrant is still valid, the IAX protocol assumes that this information is only valid for a certain time period (equal to the value enclosed in the **REFRESH IE** or the default registration value). This period is determined by the IAX registrant at the registration or registration-refresh phase.

IAX registrants are assumed to refresh their registration records before the expiration of the registration time period. From a server perspective, the IAX registrar server has to remove invalid registration records without preventing or notifying the corresponding IAX registrants.

The default registration time period is **60** seconds.

7.5.3 Retransmission

As indicated above, most IAX messages are reliable, apart from some media-specific frames (also denoted as 'mini frames'). IAX reliability is ensured owing to the exchange of explicit acknowledgement messages to notify the remote IAX peer that a given reliable IAX message has been received.

Acknowledgement messages should be sent and/or received within a certain time period (we will refer to this time period as 'retransmission period'). If no acknowledgement message has been received within the retransmission period, a retransmission of the message concerned is then required. This process is repeated until the maximum retry time period is reached.

The IAX maximum retry time period is **10** seconds.

7.5.4 Connectivity

In the context of SIP deployment, checking the aliveness of a remote party is not supported. An SIP user agent has no means of detecting that a remote peer is no longer available, and media streams may continue to be issued without interruption. Unlike SIP, IAX defines a dedicated procedure to assess the availability of a remote party.

In order to ensure that a given IAX remote participant is still alive when no audio/video frames have been received from this participant in a certain amount of time, a **PING** message has to be sent. A response must then be issued by the remote peer. This response is conveyed in a **PONG** message.

The default value of this timer is **20** seconds.

7.6 Authentication and Encryption

The intention of this section is not to detail security threats, nor to describe a generic framework for activating security mechanisms (such as setting IPSEC (Internet Protocol Security) architecture [IPSEC]), but only to describe how encryption is supported within IAX.

IAX may be used either with plain text or in conjunction with encryption mechanisms, such as AES (Advanced Encryption Standard, [AES]), which are based on a shared secret. IAX authentication is implemented by the exchange of authentication requests, which enclose **CHALLENGE IEs**. These authentication challenges should be answered by the remote peer and encrypted according to adopted encryption method. If encryption negotiation has failed, the call should be terminated. Finally, note that the authentication key used to decode the encrypted message is valid at least during the call session time.

IAX supports several authentication procedures, including:

- *Plain Text*: this method is not secure. It consists of sending the password in a clear text with no ciphering.
- *MD5*: this method is based on a password and an MD5 challenge [MD5]. This challenge is provided by the remote peer. With the password and the challenge, the receiving IAX peer computes a hash and sends it back to the remote IAX peer. Authentication is based on the hash result. Figure 7.8 provides an example of this procedure, and Figure 7.7 gives an example of IAX traces enclosing an MD5 challenge and its hash.

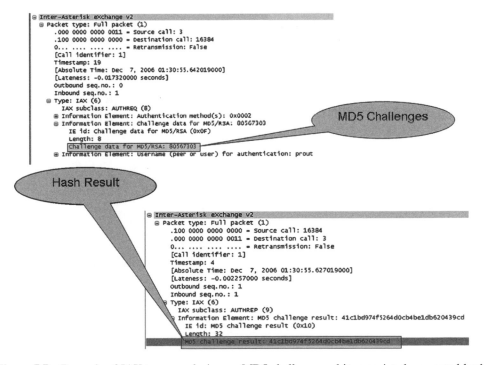

Figure 7.7 Example of IAX traces enclosing an MD5 challenge and its associated computed hash

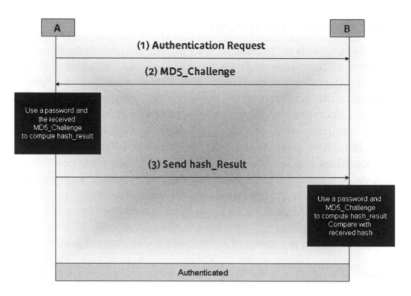

Figure 7.8   Example of MD5-based authentication

IAX does not recommend the use of this procedure and suggests RSA (Rivest, Shamir, & Adleman, [RSA]) instead. This is motivated by the fact that RSA is more robust, and is also based on the recently-discovered collisions in MD5's compression algorithm. Avoiding MD5 in new protocol specifications has been recommended.

Figure 7.9 gives an example of a security attack where MD5 is used for authentication. This figure illustrates a brute force attack. If a hacker succeeds in sniffing the traffic and retrieves the

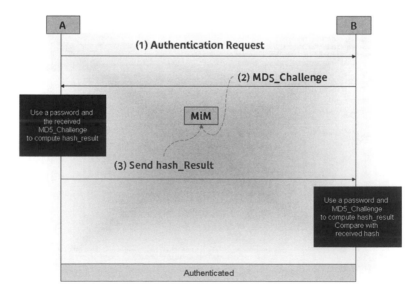

Figure 7.9   Example of MD5 attack

challenge sent by the server and the hash sent by the IAX user agent, they may detect the password. This attack assumes that the hacker (**MiM** for 'man in the middle') is in the path between the server and the user agent. Once the information is intercepted by **MiM**, a computation process is executed to compute the password. A dictionary may be used to implement this brute force attack.

7.7 Conclusion

This chapter discussed connectivity issues related to the delivery of IAX messages. It described how IAX messages are transported and reliability is implemented. IAX uses a single port number for both its signalling and media frames. These frames are transported over UDP. A dedicated reliability mechanism is supported at the application level. IAX is an IP-agnostic protocol, since no interference between service layer and network layer is introduced by IAX specification. IAX does not rely on application-specific information to initiate and manage its sessions. These sessions may be encrypted using AES or other schemes.

References

[AES] US Department of Commerce/NIST, 'FIPS-197, Announcing the Advanced Encryption Standard', November 2001

[IAX] Spencer, M., Shumard, K., Capouch, B. and Guy, E., 'IAX2: Inter-Asterisk eXchange Version 2', draft-guy-iax-04, work in progress.

[IPSEC] Kent, S. and Atkinson, R., 'Security Architecture for the Internet Protocol', RFC 2401, November 1998.

[MD5] Rivest, R., 'The MD5 Message-Digest Algorithm', RFC 1321, April 1992.

[NAT] Holdrege, M. and Srisuresh, M., 'Protocol Complications with the IP Network Address Translator', RFC 3027, January 2001.

[RFC2680] Almes, G., Kalidindi, S. and Zekauskas, M., 'A One-Way Packet Loss Metric for IPPM', RFC2680, September 1999.

[RSA] Kaliski, B. and Staddon, J., 'PKCS #1: RSA Cryptography Specifications Version 2.0', RFC 2437, October 1998.

[RTP] Schulzrinne, H., Casner, S., Frederick, R. and Jacobson, V., 'RTP: A Transport Protocol for Real-Time Applications', RFC 1889 (proposed standard), January 1996.

[SIP] Rosenberg, J., Schulzrinne, H., Camarillo, G., Johnston, A., Peterson, J., Sparks, R. et al., 'SIP: Session Initiation Protocol', RFC 3261, June 2002.

[TCP] Postel, J., 'Transmission Control Protocol', RFC793, September 1981.

[UDP] Postel, J.,'User Datagram Protocol', RFC 768, August 1980.

Further Reading

Collier, M. and Endler, D., Exploiting Voice over IP Networks', available at: http://www.hackingvoip.com/presentations/RSA%202007.pdf.

iSEC Partners, 'VoIP Security', available at: https://www.isecpartners.com/iax_brute.html.

8

IAX Operations

8.1 Introduction

This chapter focuses on IAX operations. Several call flows are provided to illustrate the behaviour of IAX protocols and the following operations are elaborated:

- *Provisioning of IAX Devices*: Section 8.2 shows how IAX supports native schemes to deliver provisioning and firmware update of IAX devices. This feature is useful in several scenarios and can avoid the activation of additional protocols and architecture.
- *Registration*: Section 8.3 details how the registration procedure is supported in IAX realms. Call flow examples are provided.
- *Call Setup*: Section 8.4 provides more examples of how a call may be initiated within IAX networks.
- *Call Tear-Down*: Section 8.5 illustrates how an active IAX session may be terminated and identifies which IAX messages should be issued to terminate an ongoing IAX session.
- *Call Monitoring*: Section 8.6 introduces IAX messages and dedicated procedures to enforce call monitoring using messages specific to IAX. No additional architectures and protocols are required to assess the aliveness of an ongoing IAX call. The procedure defined within this section bypasses a critical issue encountered by current SIP [SIP] implementations. (SIP does not provide any means to assess whether a remote peer is still alive or not. SIP messages may continue to be sent to a disconnected remote user agent (UA).)
- *Call Optimisation*: Section 8.7 defines an IAX procedure which allows a given peer to be removed from an ongoing call. This procedure allows a path-decoupled scheme.

Note that this chapter does not discuss service-specific considerations. Chapter 9 is dedicated to the support of advanced services.

Inter-Asterisk Exchange (IAX): Deployment Scenarios in SIP-Enabled Networks   Mohamed Boucadair
© 2009 John Wiley & Sons, Ltd

8.2 Provisioning and Firmware Download

8.2.1 Context

Unlike existing VoIP protocols, the IAX protocol offers native means to deliver provisioning data to a given IAX device, terminal or soft-phone. This provisioning data may be the IP address to contact in order to access the IAX service, the user name to use, the password or something of that sort. Note that the IAX provisioning messages have no assumption regarding the data carried and its encoding.

IAX provisioning is implemented mainly through a dedicated message called: **PROVISION**. Figure 8.1 provides two examples of the use of this message.

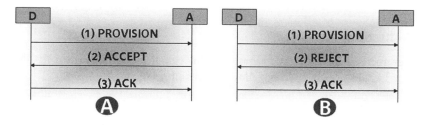

Figure 8.1   Provisioning call flow example

8.2.2 Provisioning Examples

This section provides two examples of the use of the **PROVISION** message. In these examples, **D** denotes a provisioning server and **A** an IAX client.

8.2.2.1 Successful Provisioning

Figure 8.1(A) illustrates the case where the IAX client accepts the provisioning data delivered by **D**.

Once it has detected that **A** is connected, **D** issues a **PROVISION** message destined to **A**. **A** may undertake some checking operations, such as verifying the source of the request or consulting a local access list. Once these verifications are completed, if the result is to accept the request, **A** sends an **ACCEPT** message to indicate to **D** that it accepts the provisioning data. An **ACK** message is then sent by **D**. At this stage, **A** is ready to process the provisioning data sent by **D**.

8.2.2.2 Unsuccessful Provisioning

Figure 8.1(B) illustrates the case where the provisioning request is rejected by the IAX client.

Like the previous example, once it has detected that **A** is connected, **D** issues a **PROVISION** message destined to **A**. **A** may undertake some checking operations, such as verifying the source of the request or consulting a local access list. Once these verifications are completed, if the result is to reject the request because **D** is not known to **A**, **A** sends a **REJECT** message to indicate to **D** that it rejects the provisioning data. An **ACK** message is then sent by **D**. At this stage, **A** is not ready to process the provisioning data. **D** should stop sending any data.

8.2.3 Firmware Update

8.2.3.1 Context

In addition to provisioning capabilities, IAX offers the possibility for an IAX device to check whether a new firmware version is available to be downloaded. This procedure is implemented mainly through **FWDOWNLD** and **FWDATA** messages.

- **FWDOWNLD** is used to check if there is a new firmware version.
- **FWDATA** is used by the firmware server to carry the binary block of the new available firmware version.

8.2.3.2 Call Flow Examples

Figure 8.2 provides two examples. In these examples, **A** denotes an IAX client and **D** refers to a firmware server.

Successful Firmware Update
Figure 8.2(A) illustrates the case where the firmware download request succeeds. The **FWDOWNL** request must specify the device type in order to help the server find the appropriate firmware version and check if a new one is valid. In this case, the firmware binary data is downloaded via a set of **FWDATA** message. The format of the enclosed data is device-specific and is out of the scope of the IAX protocol itself. A **FWDATA** with a **0** length **FWBLOCKDESC IE** indicates that this is the last piece of the firmware binary data. The IAX device has to issue an **ACK** message to close the session.

In this example, **A** sends a **FWDWNL** messages to the firmware server **D**. Upon receipt of this message, the server sends a response by invoking an **FWDATA** message including a block of the firmware to be updated. This process is iterated until no data is enclosed in an **FWDATA**. Then **A** sends an **ACK** message to **D** to close the firmware update session.

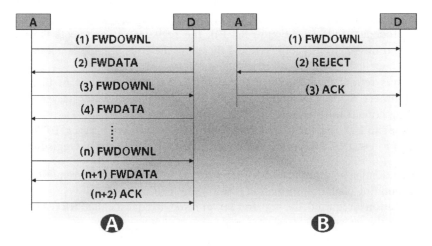

Figure 8.2 Firmware download call flow example

Unsuccessful Firmware Update

Figure 8.2(B) illustrates the scenario where no new firmware version is available for the specified device type.

In this example, **A** sends an **FWDWNL** message to the firmware server **D**. Upon receipt of this message, the server sends a negative response by invoking a **REJECT** message to reject this request. Then **A** sends an **ACK** message to **D** to close the firmware update session.

8.3 Registration

8.3.1 Procedure

The IAX registration procedure is an optional feature supported by the IAX protocol. Registration is not mandatory when the two IAX participants are able to retrieve contact information to reach each other by another means. This is only valid for a static address scheme, through a third participant or with manual configuration. If dynamic addressing is assumed, an IAX peer must register within a registrar server. This entity is responsible for maintaining IAX users' contact information (mainly location records; this entity is the same as the one used for SIP architectures). Each registrant is responsible for the validity of its contact information. IAX registrants have to issue registration requests (by invoking **REGREQ** messages) to an IAX registrar server. These requests should include information such as username and refresh timer.

- If no authentication is required, the registrar server stores the information enclosed in the **REGREQ** with the perceived connection information in a contact table. This information is returned to the registrant (see Figure 8.3(A)). The structure of the aforementioned conatct table is out of the scope of IAX protocol; nevertheless, the same structure may be used as for the current SIP registrar server. This table may be stored as a database or in a flat file.
- If authentication is required, the Registrar Server issues a **REGAUTH** request in order to challenge the registrant and assess whether the remote peer is mandated to issue such a registration request:
  - If this procedure succeeds (see Figure 8.3(B)), the registrar proceeds to store the contact information in the system (that is, instantiate a new entry in its contact table or update an existing one, if any).
  - If this procedure fails (see Figure 8.4(F)), the registrar server rejects the registration request, no information is retreived and therefore the conatct table is not updated.

IAX registrants refresh their registration records before the expiration of their associated registration timers. If no registration refresh is issued, the registrar server destroys the corresponding contact information.

At any time, an IAX registrant may deregister from the registrar server. To do so, an IAX registrant has to issue an appropriate message, called **REGREL**, to the registrar server:

- If no authentication is required, the registrar server acknowledges the request and destroys the corresponding contact information (see Figure 8.3(C)).
- If authentication is required, the registrar server issues a request called **REGAUTH** in order to challenge the registrant:
  - If this procedure succeeds (see Figure 8.3(D)), the registrar server destroys the corresponding contact information from its contact table.

- If this procedure fails (see Figure 8.4(G)), the registrar server rejects the registration release request.

8.3.2 Call Flow Examples

This section provides a set of call flows to illustrate the behaviour of IAX peers in the context of a registration procedure.

In these examples, **A** denotes an IAX registrant and **B** an IAX registrar server.

8.3.2.1 Successful Registration Call Flows

Figure 8.3 illustrates several successful registration and registration-release operations.

Successful Registration without Authentication
Figure 8.3(A) provides an example of a successful registration request when no authentication is required.

In order to connect to a VoIP service deployed using IAX, **A** must first register with the service. Therefore, **A** issues a **REGREQ** to its registrar server. The information needed to contact that registrar server may be discovered by dedicated service-discovery means such as SLP (Service Location Protocol, [SLP]), provided by a provisioning means such as that enclosed in the connectivity data in a DHCP (Dynamic Host Configuration Protocol, [DHCP]) offer, or statically configured in the user agent (UA). The registration request is then routed to the registrar server **B**.

Upon receipt of this request, **B** proceeds to some service verifications and retrieves pertinent information such as username, refresh timers and so on from the received message. This retrieved information is stored in a contact table maintained by **B**, or in an external data base. A response is then sent back to **A**. This response is a **REGACK** message, which must enclose the apparent address of **A** and, optionally, the refresh value enclosed in the received **REGREQ** message.

Once the response is received by **A**, an **ACK** message is issued from **A** to **B**.

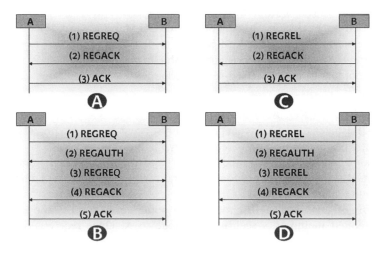

Figure 8.3   Successful registration requests

A is now registered with the service and is able to place and receive calls to/from remote destinations according to its subscribed-service contract.

Successful Registration with Authentication

Figure 8.3(B) illustrates the IAX exchanges that must occur in order to successfully process a registration request when authentication is required by the registrar.

In order to connect to a VoIP service deployed using IAX, **A** must first register to the service. Therefore, **A** issues a **REGREQ** to its registrar server. The registration request is then routed to the registrar server **B**.

Upon receipt of this request, **B** proceeds to some service verifications and issues a **REGAUTH** request including a security challenge to **A**. Once received, **A** retrieves the challenge and, with its password, computes a security hash, which is enclosed in a second **REGREQ**. This message is then sent to **B**.

Upon receipt of this request, **B** retrieves the security hash and assesses its validity. Once the hash is validated, **B** stores pertinent information such as username and refresh in a contact table or in an external data base. A response is then sent back to **A**. This response is a **REGACK** message, which must enclose the apparent address of **A** and, optionally, the refresh value enclosed in the received **REGREQ** message.

Once it is received by **A**, an **ACK** message is issued from **A** to **B**.

A is now registered with the service and is able to place and receive calls to/from remote VoIP or telephony destinations according to its subscribed-service contract.

Successful Registration Release without Authentication

Figure 8.3(C) provides an example of a successful registration release when no authentication is required.

When **A** wants to disconnect from the VoIP service, it must deregister from it. Therefore, **A** issues a **REGREL** request destined to its registrar server. The registration release request is then routed to the registrar server **B**.

Upon receipt of this request, **B** proceeds to some service verifications and retrieves pertinent information such as username from the received message. **B** destroys the records related to **A**. A response is then sent back to **A**. This response is a **REGACK** message. Once it is received by **A**, an **ACK** message is issued from **A** to **B**.

A is now deregistered from the service and unable to place or receive calls to/from remote destinations according to its subscribed-service contract.

Successful Registration Release with Authentication

Figure 8.3(D) illustrates the IAX exchanges that must occur in order to process a registration release request when authentication is required by the registrar server.

When **A** wants to disconnect from the VoIP service, it must deregister from it. Therefore, **A** issues a **REGREL** request destined to its registrar server. The registration release request is then routed to the registrar server **B**.

Upon receipt of this request, **B** proceeds to some service verifications and issues a **REGAUTH** request including a security challenge to **A**. Once received, **A** retrieves the challenge and, with its password, computes a security hash, which is enclosed in a second **REGREL**. This message is then sent to **B**.

Upon receipt of this request, **B** proceeds to some service verifications and retrieves pertinent information such as username from the received message. **B** destroys the records related to **A**. A response is then sent back to **A**. This response is a **REGACK** message. Once it is received by **A**, an **ACK** message is issued from **A** to **B**.

A is now deregistered from the service and unable to place or receive calls to/from remote destinations according to its subscribed-service contract.

8.3.2.2 Unsuccessful Registration Call Flows

Figure 8.4 illustrates several unsuccessful registration and registration-release operations.

Unsuccessful Registration without Authentication
Figure 8.4(E) provides an example of an unsuccessful registration request when no authentication is required.

In order to connect to a VoIP service deployed using IAX, **A** must first register with the service. Therefore, **A** issues a **REGREQ** destined to its registrar server **B**.

Upon receipt of this request, **B** proceeds to some service verifications. These verifications fail. A **REGREJ** message is sent back to **A**. Once received by **A**, an **ACK** message is issued from **A** to **B**.

At this stage **A** is not registered with the service and is not able to place or receive calls to/from remote destinations according to its subscribed-service contract.

Unsuccessful Registration with Authentication
Figure 8.4(F) provides an example of an unsuccessful registration request when authentication is required by the registrar.

In order to connect to a VoIP service deployed using IAX, **A** must first register with the service. Therefore, **A** issues a **REGREQ** destined to its registrar server **B**.

Upon receipt of this request, **B** proceeds to some service verifications and issues a **REGAUTH** request including a security challenge to **A**. Once received, **A** retrieves the

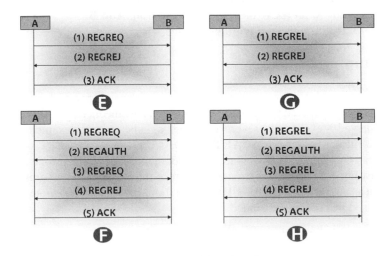

Figure 8.4   Unsuccessful registration requests

challenge and, with its password, computes a security hash, which is enclosed in a second **REGREQ**. This message is then sent to **B**.

Upon receipt of this request, **B** retrieves the security hash and assesses its validity, then proceeds to some verifications. These verifications fail. A **REGREJ** message is sent back to **A**. Once it is received by **A**, an **ACK** message is issued from **A** to **B**.

At this stage **A** is not registered with the service and is not able to place or receive calls to/ from remote destinations according to its subscribed-service contract.

Unsuccessful Registration Release without Authentication

Figure 8.4(G) provides an example of an unsuccessful registration release when no authentication is required.

When **A** wants to disconnect from the VoIP service, it must deregister from the service. Therefore, **A** issues a **REGREL** request destined to its registrar server. The registration release request is then routed to the registrar server **B**.

Upon receipt of this request, **B** proceeds to some service verifications. When these verifications fail, **B** issues a **REGREJ** message. Once it is received by **A**, an **ACK** message is issued from **A** to **B**.

At this stage **A** is still registered with the service.

Unsuccessful Registration Release with Authentication

Figure 8.4(G) provides an example of an unsuccessful registration release when authentication is required by the registrar.

When **A** wants to disconnect from the VoIP service, it must deregister from the service. Therefore, **A** issues a **REGREL** request destined to its registrar server **B**.

Upon receipt of this request, **B** proceeds to some service verifications and issues a **REGAUTH** request including a security challenge to **A**. Once received, **A** retrieves the challenge and, with its password, computes a security hash, which is enclosed in a second **REGREL**. This message is then sent to **B**.

Upon receipt of this request, **B** retrieves the security hash and assesses its validity, then proceeds to some verifications. These verifications fail. A **REGREJ** message is sent back to **A**. Once it is received by **A**, an **ACK** message is issued from **A** to **B**.

At this stage **A** is still registered with the service.

8.4 Call Setup

8.4.1 Procedure

IAX protocol is used to interconnect two peers and exchange media traffic between them. IAX call setup is achieved by exchanging several messages, as indicated in the examples in Figures 8.5 and 8.6. These messages are: **NEW, AUTHREQ, AUTHREP, ACCEPT, ANSWER** and **ACK**. Additional IAX control messages are involved, such as voice frames, for example the ones exchanged between call participants to notify them of the status of the progress of the call.

To tear down an active call, a **HANGUP** message should be sent to the remote IAX party. The call is then closed and media is stopped.

8.4.2 Successful Call Setup Flows

This section gives a brief explanation of the call setup process. Two scenarios are detailed.

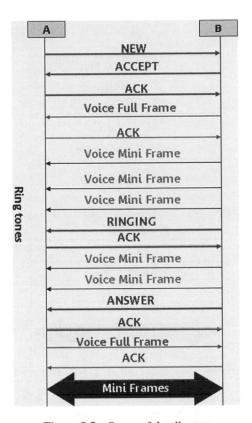

Figure 8.5   Successful call setup

8.4.2.1 Successful Call Setup without Authentication

Figure 8.5 illustrates a successful call setup and media exchange.

In order to place a call with **B**, **A** issues a **NEW** message including some appropriate information such as its name, calling context and so on.

Upon receipt of this message, **B** accepts the processing of the request and sends back an **ACCEPT** message to **A**. **A** acknowledges this message.

B sends back ring tones (conveyed in the voice mini frames just after the **ACK** message, as illustrated in Figure 8.5) and a **RINGING** message to indicate that the call is going to be answered.

When **B** sends back the **ANSWER** message, **A** acknowledges it. **A** and **B** then start to exchange media streams (mini frames).

8.4.2.2 Successful Call Setup with Authentication

This example is similar to the previous one. The only difference is that authentication is required to accept the call.

B challenges **A** by sending as a response to the **NEW** message an **AUTHREQ**. **A** has to answer the challenge via an **AUTHREP** message. If the credential included is validated by **B**, an **ACCEPT** message is sent back to **A**.

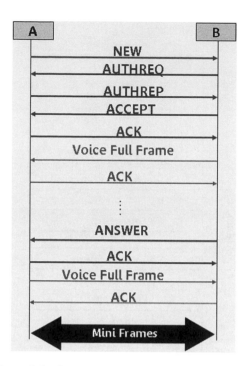

Figure 8.6   Successful call setup with authentication

B sends back ring tones (conveyed in the voice mini frames just after the **ACK** message, as illustrated in Figure 8.6) and a **RINGING** message to indicate that the call is going to be answered.

When **B** sends back the **ANSWER** message, **A** acknowledges it. **A** and **B** then start to exchange media streams (mini frames).

8.4.3 Unsuccessful Call Setup Flow

The example illustrated in Figure 8.7 shows the call flow that may be experienced when a call-initiation request is rejected by a remote peer. This is enforced by issuing a **REJECT** message in response to a **NEW** request sent by the caller party.

8.4.4 Call Setup with a Server

In the examples provided in Section 8.4.2, the calls were established directly between two IAX peers. Within an operational environment, calls are not implemented in this way. Usually, a service is hosted by dedicated servers, which are responsible for controlling access to the

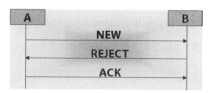

Figure 8.7   Unsuccessful call setup

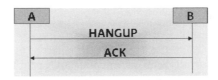

Figure 8.8 Call tear-down

service and especially for routing call requests to their destinations. Chapter 10 provides some deployment scenarios involving several IAX servers.

In the context of the involvement of IAX servers in a call, Section 8.7 provides an overview of an IAX procedure which allows an intermediary node to be removed from the call path. Messages are exchanged directly between IAX endpoints. This procedure may be suitable for service providers which want to implement a path-decoupled service.

8.5 Call Tear-Down

In order to close an ongoing IAX call, a message called **HANGUP** must be issued by one of the call participants. Figure 8.8 illustrates an example of the call flow that is experienced when **A** wants to close its session with **B**.

A and **B** are free to kill a call session whenever they want. As shown in the figure, **A** issues a **HANGUP** request, which is acknowledged by **B**. Once these two messages are exchanged, the two call participants destroy the call context.

8.6 Call Monitoring

Unlike SIP, IAX implements a method to detect whether a remote peer is still alive or not. This is achieved by an exchange of **PING/PONG** messages within a call context, or an exchange of a **POKE/PONG** outside a call context.

IAX also allows measurement of the quality of a call leg through an exchange of **PING/PONG**, **POKE/PONG** or **LAGRQ/LAGRP** messages.

A **PONG** message may encloses several information elements which assess the quality of the session. Example of these information elements are **JITTER IE**, **DROPPED FRAMES IE**, **OOO IE** (frames which are received out of order) and so on.

The value of the **Timestamp** enclosed in a **PONG** or **LAGRP** must be the same as that of the one enclosed in the initial **POK/PING/LAGRQ** request.

Figure 8.9 shows two examples of observed call flows between **A** and **C**. The first one is an exchange of **PING/PONG** messages and the second one is an exchange of **LAGREQ/LAGREP** messages.

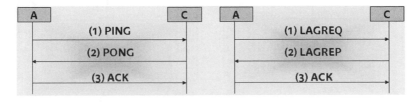

Figure 8.9 Call monitoring

8.7 Call Optimisation

8.7.1 Context

The purpose of this feature is to allow an intermediary IAX peer to remove itself from the call and let the endpoint participants communicate directly. This is desirable only if this intermediary element has no need to monitor a given call session (for example, in the context of operational deployment, accounting, billing operations and legal intercept, requirements enforce a service provider to monitor the data which is conveyed in its network).

In the context of IAX, call optimisation is achieved through an exchange of specific messages between all three participants. These messages are **TXREQ**, **TXCNT**, **TXACC** and **TXREADY**. For more information about these messages, the reader is invited to refer to Chapter 6. Note that this procedure is initiated only after an **ACCEPT** message has been sent or received for the corresponding ongoing call leg.

Section 8.7.2 describes some call flow examples to illustrate the behaviour observed during a transfer procedure.

8.7.2 Examples of a Call Transfer Call Flow

8.7.2.1 Successful Call Transfer

In Figure 8.10(A) the intermediary IAX node **B** sends two **TXREQ** messages to the remote IAX peers **A** and **C**. These messages can be sent only if the call sessions were active with both sides (that is, between **A** and **B** and between **B** and **C**). The contact information to reach the remote peer is enclosed in **TXREQ** messages.

Upon receipt of this message by **A** (respectively **C**), a **TXCNT** is directly sent to **C** (respectively **A**) in order to check the connectivity between these peers. The remote participant (**C** (respectively **A**)) answers with a **TXACC** message to indicate that the connectivity verification has succeeded.

Upon receipt of this message by **A** (respectively **C**), a **TXREADY** message is sent to **B** to indicate that the peer is ready for call transfer. Consequently, the intermediary peer **B** issues two **TXREL** messages destined to **A** and **C**.

At this stage **B** is removed from the call path, and IAX frames are exchanged directly between **A** and **C**.

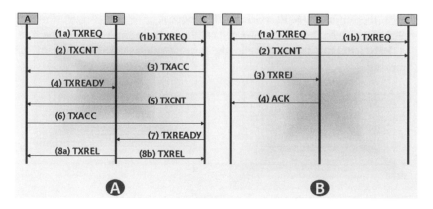

Figure 8.10   Call optimisation process

8.7.2.2 Unsuccessful Call Transfer

Figure 8.10(B) shows the case where the connectivity with the new remote peer fails. A **TXREJ** message is sent by **A** to **B**. The call transfer is consequently cancelled.

At this stage, **B** is still in the call path and all IAX messages are relayed by **B**. A call transfer may fail because of several hurdles, such as NAT or firewall rules.

8.8 Conclusion

This chapter has presented a set of examples to illustrate IAX operations. In particular, the following call flow examples have been detailed:

- Provisioning operations to highlight IAX features and capabilities to implement firmware update and provisioning.
- Registration operations responsible for handling subscription requests to the service. This procedure is important for the call-placement process.
- Call setup and tear-down operations; a set of methods related to session control and management issues. Required methods to place a new call or terminate an ongoing one are part of this functional group.
- Call monitoring operations which are supported by IAX and which aim to assess the quality of the experienced call. These methods include checking if a remote IAX endpoint is still alive or not.
- Call optimisation operations, used to implement a 'safe' path-decoupled mechanism, allowing an intermediary node to be removed from the call path. This mode is motivated by the need to ease NAT traversal and to assist endpoints in the call-establishment process and associated media frames.

Service-specific discussions are provided in Chapter 9.

References

[DHCP] Droms, R., 'Dynamic Host Configuration Protocol', RFC2131, March 1997.
[IAX] Spencer, M., Shumard, K., Capouch, B. and Guy, E., 'IAX2: Inter-Asterisk eXchange Version 2', draft-guy-iax-04, work in progress.
[SIP] Rosenberg, J., Schulzrinne, H., Camarillo, G., Johnston, A., Peterson, J., Sparks, R.et al., 'SIP: Session Initiation Protocol', RFC 3261, June 2002.
[SLP] Guttman, E.et al., 'Service Location Protocol, Version 2', RFC2608, June 1999.

Further Reading

Kempf, J., St Pierre, R. and St Pierre, P., *'Service Location Protocol for Enterprise Networks: Implementing and Deploying a Dynamic Service Finder'*, John Wiley and Sons, Ltd, 1999.
Rosenberg, J. and Schulzrinne, H., 'Session Initiation Protocol (SIP): Locating SIP Servers', RFC 3263, June 2002.
Schulzrinne, H., 'Dynamic Host Configuration Protocol (DHCP-for-IPv4) Option for Session Initiation Protocol (SIP) Servers', RFC 3361, August 2002.
Schulzrinne, H. and Volz, B., 'Dynamic Host Configuration Protocol (DHCPv6) Options for Session Initiation Protocol (SIP) Servers', RFC 3319, July 2003.

Part Two

Discussion and Analysis

Part Two focuses on some uses of the IAX protocol and on its capability to offer advanced services, to handle painful networking issues and to be easily extended so as to cover a large set of conversational features. This part SHOULD NOT be understood as THE IAX specification.

Part Two is organised as follows:

- *Chapter 9*: focuses on the support of advanced features by the IAX protocol. It looks first at the ability of IAX protocol to implement a CODEC negotiation between remote IAX peers, and at the support of 'on fly' CODEC negotiation feature. It also describes the ability of IAX to manage video sessions. A section is dedicated to an enhancement to IAX protocol which optimises the number of control messages exchanged between two remote IAX peers. The ability of IAX protocol to support 'presence services' and 'instant messaging' is analysed. Then an overview of IAX and its native support of the 'topology hiding' function and a brief overview of the support of IAX issues when 'mobile IP' is deployed are given. Finally, this chapter highlights how miscellaneous features such as call transfer, call forward, fax and so on are supported in the context of the IAX protocol.
- *Chapter 10*: is dedicated to IAX deployment in a multiservers environment. It begins by focusing on methods for the discovery of IAX resources. Two categories of these methods are listed: static and dynamic. Then an overview of the end-to-end call setup in the presence of several IAX servers in the path is provided. Moreover, load-balancing features in an IAX environment are identified and their implementation options described. There is a need for service providers to enforce path-coupled and path-decoupled architectures. The path-coupled characteristic of IAX is highlighted, and its ability to be enhanced to support a path-decoupled mode is discussed. Finally, there is a brief overview of the inability of the current IAX specification to achieve forking without avoiding routing loops, and of route-symmetry issues and the need for the signalling response path to follow the same route as the request path.

- *Chapter 11*: discusses NAT traversal issues when the IAX protocol is activated for the delivery of conversational services. The use of the IAX protocol does not introduce additional complexity to a basic IP communication. Moreover, this chapter presents the IP exhaustion problem and two ways to solve this sensitive issue for service providers. IAX can be activated in the context of those solutions. This chapter shows that IAX does not pose additional technical problems. Unlike SIP, IAX is powerful for NAT traversal and delivery of reliable communications.
- *Chapter 12*: focuses on P2P service offerings and the applicability of IAX to deliver P2P conversational services. A new architecture based on native IP capabilities is introduced. New IAX objects and messages are defined to support distributed conversational services. The proposed architecture is a multicast-based P2P architecture and does not require deployment of heavy DHT infrastructure. The proposed architecture is suitable for implementation by corporate customers since it offers flexibility and simplifies required configuration operations.
- *Chapter 13*: discusses the impact of the introduction of IPv6 on IAX-based service offerings. Several scenarios are evaluated and discussed. This chapter shows that the activation of IAX in an IPv6-enabled environment would not encounter major problems.
- *Chapter 14*: presents the notion of the 'IP telephony administrative domain' and offers a macroscopic functional view of a telephony service platform. It identifies two deployment scenarios of SBC nodes: access and interconnection. Moreover, it provides an overview of the motivations for introducing SBC nodes into SIP architectures. Two categories of motivation are identified and described: technical problems and legal requirements. A functional decomposition of an SBC node and both media and signalling considerations are given. Additionally, this chapter lists several functions as supported by SBC nodes and gives a brief overview of each one. Finally, it checks the applicability of SIP-oriented SBCs' functions in IAX-based service architectures.

9

IAX and Advanced Services

9.1 Introduction

The IAX protocol was designed initially for the delivery of audio and video telephony services. The protocol is flexible and open since it's based on the information carried in specific protocol objects called information elements (see Chapter 5). This chapter analyses the ability of IAX to support advanced services and how it can be extended to support additional features.

Chapter 9 is structured as follows:

- Section 9.2 focuses on the ability of IAX protocol to implement a CODEC negotiation between remote IAX peers. In particular, 'on fly' CODEC change is described.
- Section 9.3 describes the ability of IAX to manage video sessions, and how it manages both audio and video sessions.
- Section 9.4 is dedicated to an enhancement to the IAX protocol which optimises the number of control messages exchanged between two remote IAX peers.
- Section 9.5 assesses the ability of the IAX protocol to support 'presence services'. New messages are defined and their utilisation is illustrated.
- Section 9.6 assesses the ability of the IAX protocol to support an 'instant-messaging' service. New messages are defined and their utilisation is illustrated.
- Section 9.7 provides an overview of IAX and its native support of the 'topology hiding' function.
- Section 9.8 provides a brief overview of the support of IAX issues when 'mobile IP' is deployed. Also, considerations related to 'personal mobility' are given.
- Section 9.9 lists miscellaneous features, such as call transfer, call forward, fax and so on, and their applicability in the context of the IAX protocol.

9.2 CODEC Negotiation

IAX allows CODEC (Compression Decompression) negotiation through the support of several information elements (IEs), such as **FORMAT IE**, **CAPABILITY IE** and **CODECPREFS IE** (refer to Chapter 5 for more information). These IEs are enclosed in the first **NEW** message issued by an IAX user agent and are destined to a remote IAX user agent.

```
 Information Element: Actual codec capability: 0x0000f802
    IE id: Actual codec capability (0x08)
    Length: 4
 Actual codec capability: 0x0000f802
    .... .... .... .... .... .... ....0 = G.723.1 compression: Not supported
    .... .... .... .... .... .... ..1. = GSM compression: Supported
    .... .... .... .... .... .... .0.. = Raw mu-law data (G.711): Not supported
    .... .... .... .... .... .... 0... = Raw A-law data (G.711): Not supported
    .... .... .... .... .... ...0 .... = G.726 compression: Not supported
    .... .... .... .... .... ..0. .... = ADPCM: Not supported
    .... .... .... .... .... .0.. .... = Raw 16-bit Signed Linear (8000 Hz) PCM: Not supported
    .... .... .... .... .... 0... .... = LPC10, 180 samples/frame: Not supported
    .... .... .... .... ...0 .... .... = G.729a Audio: Not supported
    .... .... .... .... ..0. .... .... = SPEEX Audio: Not supported
    .... .... .... .... .0.. .... .... = iLBC Free compressed Audio: Not supported
    .... .... .... ...0 .... .... .... = JPEG images: Not supported
    .... .... .... ..0. .... .... .... = PNG images: Not supported
    .... .... .... .0.. .... .... .... = H.261 video: Not supported
    .... .... .... 0... .... .... .... = H.263 video: Not supported
 Information Element: CPE ADSI capability: 0x0002
 Information Element: Date/Time: Jan 22, 2006 22:04:32.000000000
```

Figure 9.1   Example of CODEC CAPABILITY IE

The remote participant must select one of the CODECs as being listed in the **NEW** message if **CAPABILITY IE** or **CODECPREF IE** has been enclosed. If **FORMAT IE** is enclosed, only the indicated CODEC needs to be used to exchange media. If the remote peer does not support the indicated CODEC, the session fails. In order to avoid such a scenario, it is recommended that **CODECPREF IE** or **CAPABILITY IE** is used to enclose more than one CODEC, as illustrated in Figure 9.1.

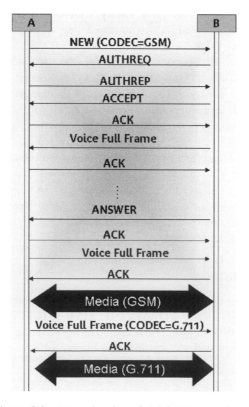

Figure 9.2   Example of 'on-fly' CODEC negotiation

In this figure, several CODECs are included in the IAX message, but only GSM (Global System Mobile) CODEC is indicated as supported.

'On fly' CODEC change can occur during the media exchange; the remote peer can use another CODEC, and indicates this in a media full frame (that is, the CODEC indicated in the last received full frame is used to transmit media flows between remote IAX user agents).

The newly-selected CODEC must be part of the **CAPABILITY IE** or **CODECPREF IE** enclosed in the first **NEW** message to be issued to initiate an IAX call.

Figure 9.2 illustrates 'on-fly' CODEC negotiation. It involves two IAX peers, **A** and **B**.

In order to place a call to **B**, **A** sends **B** a **NEW** message. This message mainly encloses **CAPABILITY IE** and a preferred CODEC positioned to GSM. **B** accepts the call request. Once the authentication procedure is completed, **A** and **B** exchange their media streams using GSM CODEC. During this call, **A** wants to change the CODEC used to G.711 [G.711], and sends a voice full frame indicating that its preferred CODEC is positioned to G.711. Once this is received by **B**, **B** sends an **ACK** message back.

At this stage, **A** and **B** exchange their media streams using the G.711 CODEC instead of the GSM one.

9.3 Video Sessions

IAX can be used to establish sessions whatever the media type is. It supports video **CODEC** negotiation and exchange of video streams. Unlike SIP, it does not allow selection of one video **CODEC** and one audio **CODEC** in the same session. Distinct IAX sessions should be set to have simultaneous audio and video streaming. These sessions will have distinct **Source Call Number** and **Destination Call Number** identifiers.

Figure 9.3 gives an example of audio and video call setups. As shown in this figure, **A** and **B** assign distinct identifiers to manage each of the ongoing media sessions. IAX control messages are exchanged in the context of each of these sessions. The type of the media session is indicated in the first **NEW** message (GSM for the audio session and H.263 [H.263] for the video one). Once control message are successfully exchanged, GSM CODEC is used to send/receive voice streams, and H.263 CODEC is used to send/receive video streams.

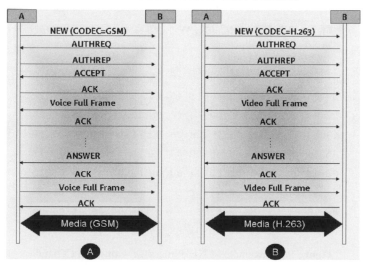

Figure 9.3 Example of audio and video call setup

The order of establishing these sessions is not important. Each IAX peer manages its ongoing sessions and distinct state machines are maintained per session.

The approach adopted by IAX to manage audio and video sessions has the advantage of managing each session type separately. An audio session may be established first, and then a video one. But this behaviour is not optimal when setting both audio and video sessions because of the number of exchanged control messages.

9.4 Negotiation of Several Media Types in the Same IAX Session

IAX allows media CODEC negotiation through the exchange of appropriate information elements, such as **FORMAT IE**, **CAPABILITY IE** and **CODECPREFS IE**. Nevertheless, only one media **CODEC** must be selected within a given IAX session. In order to exchange more than one media-type stream, several IAX sessions should be set (with distinct **Source Call Numbers** and **Destination Call Numbers**). Section 9.3 discusses the scenario of video and audio sessions.

Figures 9.4 and 9.5 provide examples of **NEW** messages that should be issued in order to set up audio (Figure 9.4) and video (Figure 9.5) sessions.

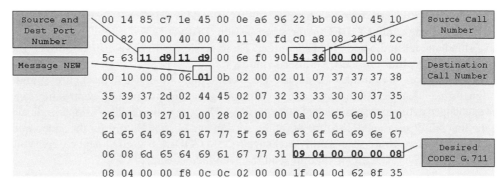

Figure 9.4 Example of **NEW** message with audio CODEC

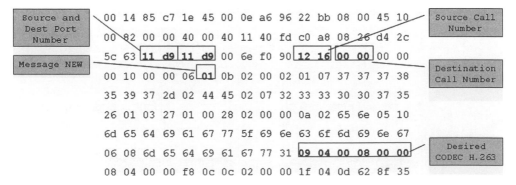

Figure 9.5 Example of **NEW** message with video CODEC

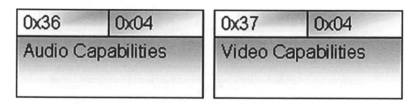

Figure 9.6   AUDIO and VIDEO CAPABILITY IEs

The **Source Call Number** assigned by the initiating IAX peer for the audio session is positioned to a value distinct from that assigned for the video session, as illustrated in Figure 9.5.

This choice seems to be nonoptimal in some scenarios (dual audio-video sessions). In order to rectify this, the IAX protocol can be enhanced so as to allow more than one media-type negotiation. A new capability is to be defined. This new capability does not make the previous behaviour obsolete; for some scenarios, the old mode may be more suitable than the new mode, and vice versa.

The proposed extension consists of allowing IAX peers to send a **FORMAT IE** per each media type as a response to a **NEW** request including **CAPABILITY IE**. In addition, this request should enclose a **CODECPREF IE** per each media type.

New information elements are introduced for this purpose, **AUDIOCAPABILITY IE** and **VIDEOCAPABILITY IE,** as illustrated in Figure 9.6.

These new information elements aim to distinguish audio and video capabilities enclosed in the media-capabilities-description part of a single IAX session.

An IAX speaker may include an **AUDIOCAPABILITY IE** and a **VIDEOCAPABILITY IE** in a single issued **NEW** message. The remote peer may select one audio **CODEC** and one video CODEC to be used during the media-exchange lifetime. It may also select one media type, and during the session lifetime an 'on fly' CODEC negotiation can occur to allow exchange of another media type.

In order to allow the assignment of preference to audio or video, it is proposed to introduce the following new information elements: **AUDIOCODECPREFS IE** and **VIDEOCODEC-PREFS IE** (see Figure 9.7).

These new information elements have the same use as the **CODECPREFS IE** but enclose either audio CODECs or video CODECs.

Thanks to the implementation of these extensions, IAX will allow exchange of audio and video streams within a single session (same **Source Call Number** and **Destination Call Number**).

Figure 9.8 shows an example of a **NEW** message enclosing both audio and video CODECs. This message indicates that the preferred audio CODEC is G.711 and the preferred video one is H.263.

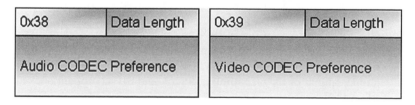

Figure 9.7   AUDIO and VIDEO CODECPREFS IEs

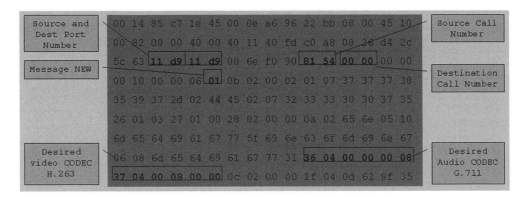

Figure 9.8 Example of **NEW** message with video and audio CODECs

Service providers should decide which implementation option is most suitable for their service architectures. For session-management purposes, the original IAX implementation option is more lightweight since it is easy to the correlate the media type and the origination session. Furthermore, since sessions are separated, no interference is to be observed between any two: if the video session breaks, the audio one may be maintained. When all media types are managed in the same session, a failure can induce unavailability of both audio and video.

9.5 Presence Services

Current specifications of the IAX protocol do not allow the implementation of presence services. Nevertheless, IAX may be easily enhanced to support them [RFC 2778].

A presence service allows users to subscribe to other users and be notified of changes in their states. This is usually associated with instant messaging. It may be implemented directly between remote peers or by an intermediary 'presence server'. [RFC 2778] defines a presence service as a set of clients: the first type of clients, called **PRESENTITIES**, provide presence information to be stored and distributed to remote peers; the second type of clients, called **WATCHERS**, receive presence information from the service. Two types of **WATCHER** may be defined: **FETCHERS** and **SUBSCRIBERS**.

- A **FETCHER** requests the current value of some **PRESENTITY**'s presence information from the presence service.
- A **SUBSCRIBER** requests notifications from the presence service of changes in some **PRESENTITY**'s presence information.

This section describes a proposal for an IAX presence service procedure close to the one defined for SIP architectures. This proposed procedure is implemented through three new messages, defined in Table 9.1. Information and data models used to convey presence information are independent of the signalling protocol, and SIP-compliant ones may be used in the context of an IAX-based presence service.

Figure 9.10 gives an example of the use of these new messages.

Table 9.1 Presence-related IAX messages

Message	Description	Enclosed IE
SUBSCRIBE	The purpose of this message is to subscribe to a presence server as an interested node in order to be notified about some events.	All information elements which may be enclosed in a **NEW** message may optionally be enclosed in a **SUBSCRIBE** message.
	A filter list about these events may be specified	An information element specifying the event filter may be enclosed
SUBSCRIBEACK	The purpose of this message is to acknowledge receipt of a **SUBSCRIBE** IAX message. Receipt of this message indicates to a remote peer to stop resending the same **SUBSCRIBE** message. If no **SUBSCRIBEACK** is received, the subscription request should be retransmitted	NA
NOTIFY	This message carries the status of the server on the occurrence of a given event. This event must be in the event filter specified by the IAX client	**STATUS** information element. The structure of this IE is as follows:

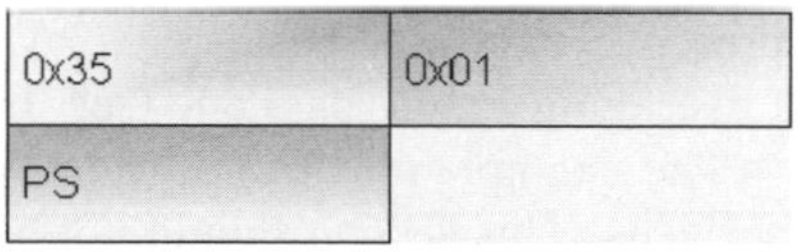

0x35	0x01
PS	

Figure 9.9 Event status IE

The **PS** field carries the status value of the remote IAX peer

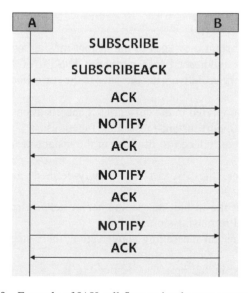

Figure 9.10 Example of IAX call flow to implement presence services

This example shows an IAX client **A** which is interested in being notified about the status of a remote IAX peer **B**. The below steps are followed to activate the presence service:

- **B** activates a presence server instance, which manages subscription requests and issues notification based on some defined events.
- To benefit from the presence service supported by **B**, **A** issues a **SUBSCRIBE** message, which may include an event filter to specify the nature of events **A** is interested in.
- **B** may refuse **A**'s subscription request and issues a **REJECT** response.
- If authentication methods are activated, **B** should issue an **AUTHREQ** and send a security challenge to **A**.
- The session won't be initiated if **A** does not provide an appropriate answer to the encryption challenge specified by **B**.
- If authentication negotiation succeeds, **B** processes **A**'s subscription request and acknowledges it by issuing a **SUBSCRIBEACK** message.
- When an event belonging to the filter specified by **A** occurs, **B** issues a **NOTIFY** message including the status of **B**.
- **A** has to acknowledge receipt of this notification.

Note that the event status may be encoded as XML-tagged data. Furthermore, the format of the presence data may be inspired by the SIMPLE (SIP for Instant Messaging and Presence Leveraging Extensions) one (see 'Further Reading').

The example provided aims to illustrate the ability of the IAX protocol to offer presence services. This section should not be understood as an IAX specification or a recommended architecture suitable for the presence service.

9.6 Instant Messaging

'Instant messaging service' denotes the ability of two peers to exchange instant messages. [RFC 2778] states that an instant-messaging service has two distinct sets of 'clients': **SENDERS** and **INSTANT INBOXES**.

- A **SENDER** provides instant messages to the instant-messaging service for delivery.
- Each instant message is addressed to a particular **INSTANT INBOX ADDRESS**, and the instant-messaging service delivers the message to the corresponding **INSTANT INBOX**.

Instant messaging is not supported in the current IAX specifications. This section proposes to enhance the protocol to support instant-messaging functions.

Two new messages are introduced to implement the instant-messaging service. These are defined in Table 9.2.

Figure 9.11 illustrates the use of the newly-defined messages in direct IAX communication:

It is recommended that the instant-messaging server acknowledges receipt of an instant message in order to avoid retransmission.

In order to initiate an instant-messaging conversation, these steps are followed:

- **A** issues a **MSG** request including a text destined to **B**.

Table 9.2 Instant messaging-related IAX requests

Message	Description	Enclosed IEs
MSG	The purpose of this message is to send a text-based message to a remote peer. The length of the enclosed data is limited by the maximum length allowed by the network	All information elements which may be enclosed in a **NEW** message may optionally be enclosed in a **MSG** message. A media IE with a subtype set to text may also be enclosed
MSGACK	The purpose of this message is to acknowledge the receipt of a **MSG** IAX message. The receipt of this message indicates to a remote peer that it should stop resending the same message. If no **MSGACK** is received, the message text should be retransmitted	NA

- **B** may refuse this message based on some local policies (e.g. black list, blocked user, etc.) or if authentication is required. In the latter case, **B** should issue an **AUTHREQ** and challenge **A**.
- The session won't be accepted if **A** does not provide an appropriate answer to the encryption challenge specified by **B**.
- If authentication negotiation succeeds, **B** processes the **MSG** and acknowledges with **MSGACK**. The purpose of this acknowledgement is to prevent retransmission of the same text message.

When an instant-messaging server is deployed to relay the instant-messaging messages, the exchange shown in Figure 9.12 occurs.

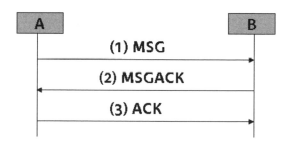

Figure 9.11 IAX messages exchange for direct IM services

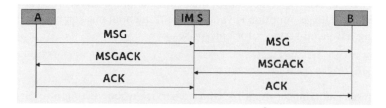

Figure 9.12 IAX messages exchange for instant-messaging services

In this example, **IMS** (Instant-Messaging Server) intercepts instant messages issued by **A** and **B** and delivers them to their final destinations. **IMS** may implement some control policies to control unwanted traffic and to prevent against spam attacks. This configuration is suitable for a provider-provisioned instant-messaging service.

9.7 Topology Hiding

IAX does not carry any information about the service architecture topology, nor about the crossed intermediary IAX nodes. Consequently the 'topology hiding' (commonly referred to as 'THIG') function (at the service layer) becomes obsolete within IAX architectures.

This is one of the major differences between IAX and SIP, since SIP messages carry information about the crossed services nodes, which may be critical in operational deployments of conversational services. Within SIP architectures, service providers should insert intermediary nodes, called SBCs (Session Border Controllers, [SBC]), in order to enforce THIG function and delete information about the topology of the service platform. This is because SIP conveys critical information related to the internal service topology, mainly in **Route**, **Record-Route** and **Via** headers. These headers are modified or even deleted before forwarding the SIP message to the next hop or final destination. This function is commonly implemented at access nodes (those elements used to contact customers' equipment with the service platform) or interconnection elements (used to interconnect a service provider's domain with external realms managed by other service providers).

IAX does not expose data related to service topology, nor the identities of IAX servers that have been involved in the service-delivery chain.

9.8 Mobility

The purpose of this section is not to explain how to implement mobility nor to detail mobile IP protocols, but only to provide some recommendations regarding how to use IAX to furnish mobility in all its facets.

9.8.1 Personal Mobility

The current IAX specification allows the registration of one and only one IP location. This restriction can be relaxed and the protocol enhanced to allow registration of the contact information of several IPs. Consequently, a given subscriber can use several devices, several contexts, several IP addresses to connect to the service. A preference may be enforced to characterise all these locations and an IAX server may be enhanced to implement this service logic.

Note that the restriction to a single IP is not motivated in the IAX specifications. It seems to be a technical decision to avoid specification of the behaviour of the server when more than one IP contact is available. This assumption is valid for the residential market, but becomes arguable when deploying advanced services for enterprises.

9.8.2 IP Mobility

A mobile terminal can connect to an IAX-based service when visiting a foreign network. In order to avoid changing the registered IP address, it is recommended that the home address and not the care-of address be registered. In such a scenario, the same IP address will be used to issue the IAX requests/responses. Media messages will be delivered to the mobile terminal even when there is a change of visitor network.

The continuity of the service is ensured by mobile IP nodes and not service elements, due to the existence of a tunnel between HA (Home Agent) and FA (Foreign Agent) entities.

Further studies of media optimisation and maintenance of media exchange should be undertaken. These studies are out of the scope of this book.

9.9 Miscellaneous

Traditional conversational services are coupled with a set of high-level features, such as the ones listed below:

- *Call Forward*: this advanced feature consists of transferring a received call to a predefined number. This feature can be implemented by the IAX protocol using the **TRANSFER** information element.
- Call forward is therefore natively supported by IAX.
- *Call Line Identification/Number Presentation*: these features aim to provide information about the calling name, calling number and so on. IAX natively supports these features through the **CALLINGNUMBER** and **CALLINGNAME** informational elements.
- CLIR (Calling Line Identity Restriction) and CLID (Calling Line Identification) features are therefore natively supported by IAX.
- *Message Waiting*: this advanced service consists of informing the customer that new messages have been received. This feature can be implemented by the IAX protocol using the **MWI** IAX request.
- Message waiting is natively supported by IAX.
- *DTMF*: IAX supports DTMF through the invocation of a **DIAL** request.
- DTMF is natively supported by IAX.
- *Music on Hold*: this is natively supported by the IAX protocol by invoking a **HOLD** request and a **MUSICONHOLD** informational element.
- Music on hold is natively supported by IAX.
- *Fax*: iaxmodem [IAXFAX] is an IAX module that can be used to send fax data. No specification document is available.

9.10 Conclusion

Unlike SIP, IAX natively offers a large set of PSTN (Public Switched Telephone Network) services. In IAX, for instance, there are no issues related to the information used for number presentation such as SIP has. Indeed, the IAX protocol natively offers classic telephony services through the use of information elements. Nevertheless, some advanced features such as presence service and instant Messaging are not supported natively by the core specification of the IAX protocol. These advanced services are not supported by the core specifications of the SIP protocol (RFC3261) either. IAX is easily extended and evolved to offer advanced services.

Nevertheless, protocol designers and service providers should decide whether a single-protocol approach is a good approach, or if simple and lightweight protocols are preferable. The simplicity of the protocol is driven by its provided functions and initial requirements. IAX was initially designed to deliver audio/video telephony services. It can easily be extended to deliver other types of service. But the richness and flexibility of the IAX protocol should be balanced by its complexity.

References

[DTME] Schulzrinne, H. and Petrack, S., 'RTP Payload for DTMF Digits, Telephony Tones and Telephony Signals', RFC2833, May 2000.

[G.711] ITU-T Recommendation G.711, 'Pulse Code Modulation (PCM) of Voice Frequencies', International Telecommunication Union (ITU-T), November 1988.

[RFC 2778] Day, M., Rosenberg, J., and Sugano, H., 'A Model for Presence and Instant Messaging', RFC 2778, February 2000.

[H.263] ITU-T Recommendation H.263, 'Video Coding for Low Bit Rate Communication', International Telecommunication Union (ITU-T), July 1995.

[IAXFAX] https://sourceforge.net/projects/iaxmodem.

[SBC] Hautakorpi, J. et al., 'Requirements from SIP (Session Initiation Protocol) Session Border Control Deployments', draft-camarillo-sipping-sbc-funcs-05.

[SIP] Rosenberg, J., Schulzrinne, H., Camarillo, G., Johnston, A., Peterson, J., Sparks, R. et al., 'SIP: Session Initiation Protocol', RFC 3261, June 2002.

Further Reading

Open Mobile Alliance, 'OMA Instant Message and Presence Service', V1.1, November 2002.

Open Mobile Alliance, 'OMA Presence Simple V1.0', July 2006.

SIMPLE Working Group, http://www.ietf.org/html.charters/simple-charter.html.

3GPP TR 23.841 (technical report), 'Presence Service: Architecture and Functional Description'.

3GPP TS 23.141 (technical specification), 'Presence Service: Architecture and Functional Description: Stage 2'.

3GPP TS 24.141 (technical specification), 'Presence Service Using the IP Multimedia (IM) Core Network (CN) Subsystem: Stage 3'.

Peterson, J.,'Address Resolution for Instant Messaging and Presence', RFC 3861, August 2004.

Saint-Andre, P.,'Extensible Messaging and Presence Protocol (XMPP): Core', RFC 3920, October 2004.

Saint-Andre, P.,'End-to-End Object Encryption in the Extensible Messaging and Presence Protocol (XMPP)', RFC 3923, October 2004.

Saint-Andre, P.,'Extensible Messaging and Presence Protocol (XMPP): Instant Messaging and Presence', RFC 3921, October 2004.

10

Multi-IAX Servers Environment

10.1 Introduction

Multimedia (e.g. telephony, video, etc.) service offerings are delivered to end users by an orchestration of several service nodes enforced by the service provider (through engineering rules, policies and so on). The service logic may be hosted in a single or centralised platform or be distributed among several portions of the service nodes. From this perspective, the introduction of IAX into operational networks should meet several requirements expressed by service providers. This chapter focuses on the ability of IAX to be deployed in a multiserver environment.

Chapter 10 is structured as follows:

- Section 10.3 focuses on methods to enforce discovery of IAX resources. Two categories are listed: static and dynamic.
- Section 10.4 provides an overview of the end-to-end call setup in the presence of several IAX servers in the path.
- Section 10.5 is dedicated to implementing load-balancing features in an IAX environment. Both static and dynamic approaches are presented.
- Section 10.6 discuses the need for service providers to enforce path-coupled and path-decoupled architectures. This section describes the current practices with regard SIP. The path-coupled characteristic of IAX and its ability to be enhanced to support a path-decoupled mode are then discussed.
- Section 10.7 provides a brief overview of the inability of the current IAX specification to achieve forking without avoiding routing loops.
- Section 10.8 is dedicated to route symmetry issues and the need for the signalling response path to follow the same route as the request path.

10.2 Focus

The IAX protocol can be used to set up one or more call legs within an end-to-end call, just like VoIP signalling protocols. In this chapter we focus on IAX in this role.

Inter-Asterisk Exchange (IAX): Deployment Scenarios in SIP-Enabled Networks   Mohamed Boucadair
© 2009 John Wiley & Sons, Ltd

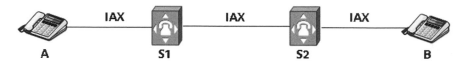

Figure 10.1   Multi-IAX servers environment

Figure 10.1 illustrates a scenario in which two IAX clients are associated with distinct IAX servers. **A** and **B** are identified by their IAX URIs.

This chapter focuses on the following:

- Discovery of IAX resources.
- Setting end-to-end IAX calls.
- Load balancing.
- Decoupling the media path from the signalling one.

10.3 Discovery of IAX Resources

Numerous methods may be implemented to discover IAX resources. This section details some of the methods available to IAX speakers to discover the IP information needed to contact a remote IAX peer.

10.3.1 Static Configuration

IAX servers may be configured statically (manually or not) in order to nominate an IP address which is to be contacted to deliver calls to telephony prefixes. These prefixes are IAX compliant, as described in Chapter 3.

In order to set up a call to prefixes managed by **S2**, **S1** should be configured to route IAX calls to **S2** (if they are destined to a peer of its prefixes).

When **A** issues an IAX request to **B**, it is sent to **S1**. Upon receipt of this request, **S1** forwards it to **S2**. The request is then forwarded to **B**. The call can then be established between **A** and **B**.

10.3.2 DUNDi

10.3.2.1 Overview

DUNDi (Distributed Universal Number Discovery, [DUNDI]) is a distributed and decentralised system, composed of a collection of peers (denoted DUNDi peers), suitable for locating telephony resources in the context of telephony delivery services. DUNDi does not rely on a centralised authority nor on a centralised node. In this way, no single point of failure is present in the system when DUNDi is activated. DUNDi is not a signalling protocol to set up calls. It is used to locate a given resource in the context of call placement. It can be used jointly with other signalling protocols such as IAX, SIP, or H.323. Exchanges between the aforementioned DUNDi peers can be secured mainly by AES (Advanced Encryption Standard, [AES]) for encryption and RSA (Rivest, Shamir, & Adleman, [RSA]) for authenticating information.

Within a DUNDi system, each node is identified by a globally-unique entity identifier, whose value typically is an Ethernet MAC (Media Access Control) address for one of the network interfaces of the system. Example of these identifiers are listed below (an example of a DUNDi map can be found at: dundi-map.netmonks.ca/):

- **00:02:b3:1c:eb:18** owned by Portafone/Germany.
- **00:08:a1:4c:82:03** owned by ISP Service Ltd.
- **00:0b:cd:ca:27:50** owned by Free World Dialup.
- **00:0c:76:8c:59:1c** owned by SpeakUp Peering.
- **00:01:02:14:ee:b9** owned by StuStaNET.
- **00:40:f4:52:1d:4a** owned by Voiceflux.com.

Within a DUNDi system, each node is connected to at least one other DUNDi peer. For robustness and availability reasons, it is recommended that each DUNDi node be connected with at least two other nodes in the system. To locate a resource within a DUNDi system, a given peer issues a request to its adjacent peers. If one of these adjacent peers finds a response, it sends a response in turn to the originating peer. Otherwise, each adjacent peer forwards the request in its turn to its adjacent peers. This procedure is iterated until the targeted resource is located.

DUNDi uses a **DPDISCOVER** message to issue a resource-location request and **DPRE-SPONSE** to send a response. Prior to issuing a **DPDISCOVER** request, a given node must register on the system by invoking a **REGREQ** message. This is acknowledged by a **REGRESPONSE** message. Each registration is conditioned by an expire timer indicated in the issued registration request.

DUNDi specifications enclose methods to optimise the queries and to optimise the lookup delays.

10.3.2.2 Discovery of IAX Resources Using DUNDi

The DUNDi protocol can be activated between **S1** and **S2**. **S1** and **S2** can collaborate in order to determine the server which is responsible for handling a given telephone extension or IAX prefixes. Note that DUNDi can be set between a large number of peers, leading to building a trusted network.

Let's suppose that **A** wants to place a call destined to **B**. Prior to establishing a call between **A** and **B**, **S1** and **S2** are configured to activate DUNDi between themselves. Thus, **S1** (respectively **S2**) sends a **REGREQ** message to **S2** (respectively **S1**), as illustrated in Figure 10.2.

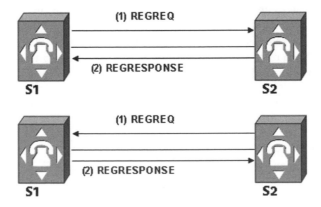

Figure 10.2 DUNDi registration

In order to place a call between **A** and **B**, the steps below are followed (Figure 10.3):

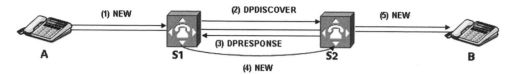

Figure 10.3   Example of call establishment (DUNDi)

• **A** issues a **NEW** request to **S1, because S1** is configured as its outbound IAX proxy server.
• Upon receipt of this request, **S1** checks its local contact table. **B** is not registered in this table, so **S1** issues a DUNDi request in order to determine the server managing the **B** IAX resource.
• Because **S2** is a DUNDi peer of **S1**, it answers positively to **S1**, notifying it that it is responsible for managing **B**.
• **S1** forwards the IAX request to **S2**.
• Upon receipt of the request, **S2** checks its contact table for **B**. The request is then forwarded to the AoC (Address of Contact) of **B**.

10.3.3 TRIP

10.3.3.1 Overview

TRIP (Telephony Routing over IP, [TRIP]) is a routing protocol that can be enabled within a single telephony domain managed by a single administrative entity, or between two telephony domains. TRIP can be used independently of the telephony signalling protocol. TRIP design is inspired by the Border Gateway Protocol (BGP) and enhanced by some link-state features, such as the Open Shortest Path First (OSPF), IS-IS and the Server Cache Synchronization Protocol (SCSP).

The main function of a TRIP speaker, also denoted 'location server' (LS), is to exchange information with adjacent LSs. This information includes the reachability of telephony destinations and the routes to these destinations. TRIP can be used to exchange attributes in order to enforce local policies, and to select routes based on these policies.

Each LS maintains a set of telephony routing tables to store local and learned reachability information. These tables are used for lookup operations driven by an external proxy server. Note that TRIP can manipulate several telephony URIs, such as SIP and H.323.

Figure 10.4 provides an example of a topology composed of five IP telephony administrative domains (ITADs). Each of these ITADs deploys several LSs. These LSs manipulate several types of route, stored in distinct telephony routing information bases (TRIB) as shown in Figure 10.5.

• **Adj-TRIBs-In (Adjacent TRIB In)**: stores routing information conveyed in TRIP messages received from adjacent LSs. These routes are used as input to the telephony routing decision process. A given LS maintains an **Adj-TRIB-In** per adjacent LS.
• **Ext-TRIB** (External TRIB): only one **Ext-TRIB** is maintained by a given LS. This table contains the results of the routing selection process applied to both local and external routes. Only one route is selected per given telephony destination.

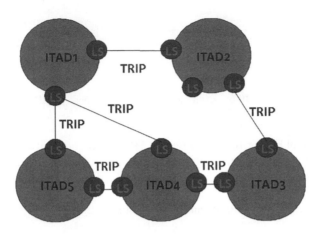

Figure 10.4   Example of TRIP topology

- **Loc-TRIB** (Local TRIB): this table stores local routes.
- **Adj-TRIB-Out** (Adjacent TRIB Out): This table stores routes which are advertised by a local LS to its adjacent LSs.

Interconnection agreements are settled between adjacent domains (Figure 10.4). As a consequence, two adjacent LSs exchange their telephony routing by the activation of TRIP. Each of these ITADs is therefore aware of available telephony prefixes which are reachable through adjacent domains. The establishment of telephony sessions is achieved using a dedicated signalling protocol (e.g. SIP or IAX).

10.3.3.2 Discovery of IAX Resources Using TRIP

This section suggests the use of TRIP as the routing protocol to convey IAX resources. Below are some operational details about the use of IAX jointly with TRIP. This section assumes that TRIP should be extended to include IAX URI, since it currently only supports SIP and H.323 URIs.

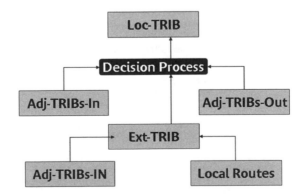

Figure 10.5   Telephony routing information tables

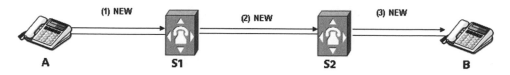

Figure 10.6   Example of call establishment (TRIP)

Within the context of Figure 10.1, suppose that TRIP is activated between **S1** and **S2**. **S1** and **S2** exchange their telephony routing tables. Consequently, **S1** will own a telephony route to reach **B**. The TRIP session between **S1** and **S2** is maintained through regular exchange of TRIP-specific messages.

In order to place a call between **A** and **B**, these steps are followed (Figure 10.6):

- **A** issues a **NEW** request to **S1**, because **S1** is configured as its outbound IAX proxy server.
- Upon receipt of this request, **S1** checks its local contact table. Since **S1** has received the telephony routes of **S2**, **B** is found in the TRIB (Telephony Routing Information Base) of **S1**.
- **S1** forwards the IAX request to **S2** (since it is indicated a **NEXT HOP**). An alternative option is to send the request directly to **B**.
- Upon receipt of the request, **S2** checks its contact table for **B**. The request is then forwarded to the AoC (Address of Contact) of **B** and the call can be established between **B** and **A**.

As illustrated in Figure 10.6, the messages required to place a call between **A** and **B** when there are several IAX servers in the call path are optimal with TRIP compared to with DUNDi. A cache may be activated to enhance required call setup delay.

10.4 Setting End-to-End Calls

In all the aforementioned methods to retrieve the location of **B**, the call flow in Figure 10.7 is observed. We assume that no authentication scheme is activated between two adjacent IAX speakers. To simplify the call flow, the provided example does not include voice frame exchanges.

As illustrated in Figure 10.7, the steps below are followed:

- **A** sends its **NEW** request to **S1** so as to request a call establishment with **B**.
- Once received by **S1**, it checks its location table and invokes a pertinent method to retrieve the location of **B**. As a result, the AoC of **S2** is found.
- The **NEW** request is then forwarded to **S2**.
- A **PROCEEDING** message is also sent to **A**, to notify it that the request is currently being handled by **S1**.
- In its turn, once the request is received, **S2** consults its location table and retrieves the AoC of **B**. Then the request is forwarded to the AoC of **B**.
- A **PROCEEDING** message is also sent to **S1**, to notify it that the request is currently being handled by **S2**.
- Once the request is received, **B** accepts the call and issues an **ACCEPT** message.
- This message is relayed by **S2** and **S1** until delivered to **A**.

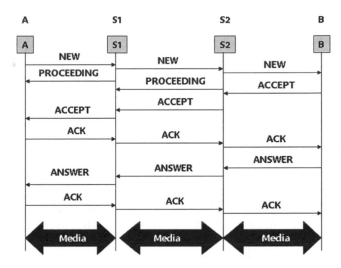

Figure 10.7   Multi-IAX servers environment call flow example

- When the **ACCEPT** message is received, **A** sends back an **ACK** message, which is relayed by **S1** and **S2**.
- As a final response to this **ACCEPT** message, **B** sends an **ANSWER** message to accept the call establishment.
- An **ACK** message is sent by **A** to acknowledge the receipt of that **ANSWER** message.
- Media frames can be then exchanged by **A** and **B**. These media frames are relayed by **S1** and **S2**.

Figure 10.8 illustrates the IAX message exchange that must occur in order to tear down an established session.

As illustrated in Figure 10.8, the steps below are followed:

- **A** sends its **HANGUP** request to **S1** so as to request a call tear-down.
- Once it is received, **S1** sends the **HANGUP** request to **S2**.
- A **PROCEEDING** message is also sent to **A** to notify it that the request is currently being handled by **S1**.

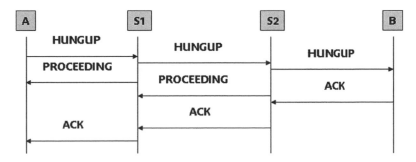

Figure 10.8   Multi-IAX servers environment tear-down example

- In its turn, once the request is received, **S2** forwards it to the AoC of **B**.
- A **PROCEEDING** message is also sent to **S1** to notify it that the request is currently being handled by **S2**.
- Once the **HANGUP** is received, **B** accepts and issues an **ACK** message.
- The call is then terminated between **A** and **B**.

10.5 Load Balancing

10.5.1 Objective

VoIP service providers may require deployment of load-balancing techniques in order to place calls. This type of load balancing is closely associated with the underlying telephony routing system and not with signalling protocols. Nevertheless, this section provides some proposals to offer load balancing within an IAX environment. Load balancing is the capability to distribute the load between several servers and connection links.

Load balancing can be implemented in conjunction with the IAX protocol. The reference architecture to conduct this analysis is illustrated in Figure 10.9:

10.5.2 Implementation Alternatives

Two approaches are elaborated below: the static and the dynamic procedure.

10.5.2.1 Static Procedure

A static procedure may be implemented. **S1** can be configured to route a certain number of calls to **S2** and another number to 'S3'. Of course, **S1** must be sure that **S2** and **S3** own routes to reach a given set of telephony destinations.

Static configuration is suitable with multi-homing but does not optimise the use of available routes to reach a given set of remote telephony destination. This motivation has justified the introduction of the dynamic scheme detailed below.

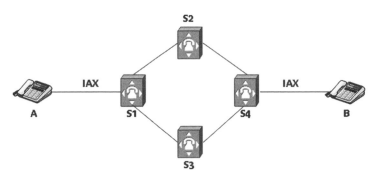

Figure 10.9 Load balancing

10.5.2.2 Dynamic Procedure

As indicated above, DUNDi or TRIP can be used to share telephony number locations and therefore to build a trusted infrastructure. These protocols may be enhanced in order to convey information about the status of the load of the overall discovered route (e.g. 50% of unused resources, 15% of unused resources, etc.). This information is updated in a cascaded way during the discovery phase. These extensions can be exploited by any VoIP signalling protocol, such as IAX. The meaning of the exchanged load information is globally shared between all peers.

An enhanced DUNDi mode is elaborated below.

Let us suppose that DUNDi is activated between **S1** and **S2**, **S1** and **S3**, **S3** and **S4** and **S2** and **S4**, as illustrated in Figure 10.10.

In this context, an enhanced mode with load balancing is illustrated in Figure 10.11.

In order to set up a call between **A** and **B**, the steps below are followed:

- **A** sends a request to place a call with **B**. This request is sent to **S1**, since **S1** is **A**'s service contact.
- When the request is received, **S1** issues a DUNDi request to its DUNDi peers enquiring whether they have a valid route to **B**.
- **S2** and **S3** do not know **B**; they issue new DUNDi requests, which are received by **S4**.
- **S4** knows **B** and it responds to **S2** and **S3**. **S4** encloses an indicator to assess its load. This indicator is denoted **load_S4**.
- Upon reception of this response, **S2** (respectively **S3**) updates it with its local load status (**load_S2** (respectively **load_S3**)) and transmits it to **S1**.
- **S1** has at its disposal two routes to reach **B**. **S1** may decide to equilibrate its requests between **S2** and **S3** or to select the less loaded server based on received information and so on. In this

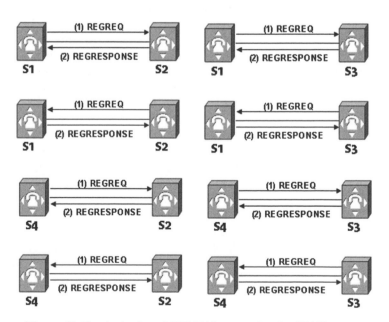

Figure 10.10 Activation of DUNDi between involved IAX servers

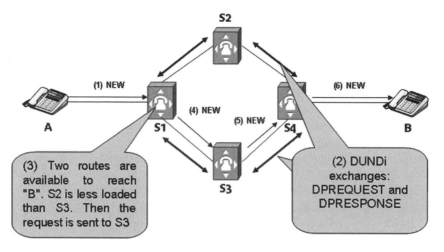

Figure 10.11   Activation of enhanced DUNDi between involved IAX servers

example, the second option is taken by **S1**. Since **S3** is less loaded than **S2, S1**forwards its request to **S3**.

• The request is forwarded to **S4,** which in turn relays it to **B**.

10.6 Path-Coupled and Path-Decoupled Discussion

10.6.1 Service Provider Requirements

In current VoIP service architectures, service providers need to control the traffic which enters into their operational networks. They also need to correlate the media traffic with the signalling traffic which has led to that media exchange session. This is motivated mainly by the need to issue CDR data (Call Data Records), used for accounting and billing purposes. Furthermore, in order to meet legal requirements such as legal intercept and emergency calls, the traffic should be 'trapped' and controlled before accessing core service nodes. Beside these legal constraints, service providers need to hide their service topology and prevent it from being exposed to external parties.

In deploying their service offerings, at least two types of protocol may be considered:

• *Path-Decoupled Protocols*: characterised by the separation between the nodes that treat the signalling data and the ones that will be invoked to treat the resulting media streams.
• *Path-Coupled Protocols*: a set of protocols which are designed in such a way that the media path is the same as the signalling one.

Service traffic is aggregated in access nodes (usually denoted as PoP (Point of Presence)) and then relayed to core service elements. These elements may be centralised or else organised into several platforms. For billing and SLA assurance and fulfilment purposes, all multimedia sessions must be traced. Therefore, core service nodes must be stateful (maintain states related to ongoing sessions) and not stateless.

Inside an IP telephony administrative domain (ITAD), the path followed by the first signalling message within the context of a given session should be used by all signalling messages of that session unless all involved service elements are synchronised or a centralised element is deployed to maintain ongoing sessions.

10.6.2 SIP: a Path-Decoupled Protocol

10.6.2.1 Theory

SIP is a flexible protocol that has been designed to be decoupled. SIP decouples the signalling path from the media one using its negotiation capabilities feature. The SIP signalling path can be distinct from the media path according to the SDP (Session Description Protocol) offers and answers. The signalling path may be stored in a specific header called **Via** or **Record-Route**. The media path is end-to-end, and endpoints are indicated in the SDP part.

This flexibility of SIP has introduced several technical problems, such as NAT and firewall traversal (e.g. due to dynamic assignment scheme of the port used to send/receive RTP (Real-Time Transport Protocol) flows). Additional applications are required to modify SIP messages and help the crossing of intermediary nodes.

In order to provide stateful call features, SIP messages enclose specific headers such as **Via**, **Route** and **Record-Route** so as to guide and drive routing of the issued signalling messages and their associated responses. Therefore, the signalling path of a request may be forced by **Route** headers, and that of the response follows what is indicated in **Via** and/or **Record-Route**.

10.6.2.2 Practices

The aforementioned flexibility of the SIP protocol is suitable within a VoIP service provider realm but not at the access segment, since VoIP service providers have to insert intermediary nodes in order to achieve some specific functions, such as shaping and policing, legal interception, emergency calls and so on.

To do so, VoIP service providers deploy SIP at the access segment of their service offerings architectures but within a path-coupled scheme. These inserted intermediary nodes break the end-to-end paradigm of the SIP signalling protocol and introduce SIP-unfriendly behaviours. The reader is invited to refer to [SBC] for more information about the incoherence induced by such nodes.

To enforce stateful behaviours, service providers use **Via**, **Route** and **Record-Route** headers that are provisioned with appropriate information to drive the signalling path. The media path is enforced by the involvement of intermediary nodes such as SBCs. These SBCs act as back-to-back user agents.

Chapter 14 provides more information regarding SBC nodes.

10.6.3 IAX: a Path-Coupled Signalling Protocol

Unlike SIP, IAX adopts a pragmatic approach which takes into account the presence of intermediary nodes such as NAT boxes. The main preoccupation of the IAX protocol is to establish a successful session between call participants, rather than to be flexible. Therefore, only IAX servers that received an IAX message from a given IAX client can contact this IAX client. All inbound calls destined to this IAX client should be set via these servers (of course, a direct IAX request may be sent to this client but there is no assurance that the call would be successfully established, due to NAT problems).

Once the session has been set up, a given intermediary server may remove itself from it and let the call participants exchange IAX data directly between them. This process is conditional on the acceptance of both parties and the success of direct message exchange between the call participants.

Within a path-decoupled scheme, for instance SIP, some sessions may not be established even when an intermediary server is deployed (both signalling and media ports should be opened in the NAT box, for instance). But all IAX sessions will be established using IAX.

10.6.4 Discussion

10.6.4.1 Basic Procedure

In some scenarios, especially in those requiring an optimisation of the media path, a path-decoupled scheme should be adopted (e.g. inside a VoIP service provider realm).

Within this section, the reference architecture shown in Figure 10.12 is adopted.

For instance, this architecture may be adopted for IAX operational service in:

- 'Small' IAX servers which are associated with each access POP.
- A central IAX server which is responsible for some specific functions such as user profile management, authentication and so on.

This example is not a recommendation of the service configuration and setup, but only an exercise to check the ability of the IAX protocol to offer direct media communications between access nodes. By access nodes, we mean IAX servers located at the edge of an IAX-based telephony domain.

This elaborated procedure follows the classical IAX behaviour. After establishment of the IAX session, more precisely after exchanging **ACCEPT** messages between involved IAX parties (Figure 10.7 illustrates the procedure to set up the call between **A** and **B**), **S2** can remove itself from the path.

Figure 10.13 reproduces message exchanges involving **S1**, **S2** and **S3**. This procedure will always succeed since these IAX nodes are managed by the same administrative entity and no complications related to NAT traversal or firewall policies should be experienced.

As shown in this figure, once the media flows are exchanged between **A** and **B**, one of the intermediary IAX servers may be removed from the media path so as to optimise end-to-end delay and not load all involved servers. In this example, **S1** and **S3** are assumed to be IAX

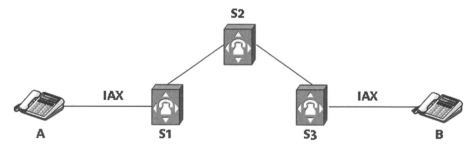

Figure 10.12   Path-decoupled scenario

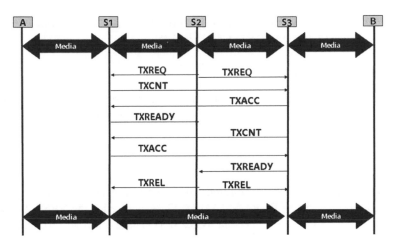

Figure 10.13   Path-decoupled call

servers deployed at the access segment, and **S2** an IAX server located at the core service platform. The service logic is principally enforced by **S2**. Once access to the service is granted and the call has been placed between **A** and **B**, **S2** may be removed from the path, leaving only **S1** and **S3** to relay the media path. **S1** and **S3** may also be removed, and the media flows will be exchanged directly between **A** and **B**. In this example, we don't detail the latter scenario; we assume that a given service provider needs to control the session and that its service nodes should be maintained in the path.

In order to enforce a path-decoupled scheme, the steps below are followed:

- **S2** issues two **TXREQ** messages to initiate a call transfer and to be removed from the media path. These requests are sent to **S1** and **S3**. During this transfer operation, an identifier is assigned by **S2** to uniquely identify the ongoing session. This identifier is maintained during the transfer operation. **TXREQ** messages include the IP address and port number of the remote peer to which the call should be transferred (in this case the IP address and a port number of **S1** and **S3**).
- Once received by **S1** (respectively **S3**), a **TXCNT** is sent to **S3** (respectively **S1**) to check that no problems are encountered in reaching **S3** (respectively **S1**). The **TXCNT** message is used to verify the connectivity with the new peer to which the call should be transferred. The identifier enclosed in the received **TXREQ** is used when issuing the **TXCNT**.
- **S3** (respectively **S1**) sends a **TXACC** message, to acknowledge the receipt of the **TXNT** message, to **S1** (respectively **S3**). This message is a response to a **TXCNT** request, issued in order to accept to transfer request. The same transfer identifier sent in the **TXCNT** is enclosed in this message.
- A **TXREADY** is sent by **S1** and **S3** to **S2** to notify it that the transfer procedure has succeeded. **S2** stops sending media to the former location.
- At this stage, **S2** sends **TXREL** messages to **S1** and **S3** to terminate the transfer process.
- When the **TXREL** message is received by **S1** and **S3**, media flows (IAX mini frames or video full frames) are exchanged directly between **S1** and **S3**. The media path is then modified and does not include **S2**.

Note that during the call optimisation phase, media flows (IAX mini frames or video full frames) are exchanged between **A**, **S1**, **S2**, **S3** and **B**.

10.6.4.2 Optimised Procedures

Since **S1**, **S2** and **S3** are managed by the same administrative entity, the behaviour of the IAX protocol with regard the transfer process may be enhanced so as to reduce the number of IAX messages exchanged during a transfer operation.

10.6.4.3 First Alternative

A first alternative to enforce a path-decoupled scheme is: during the call establishment process, illustrated in Figure 10.7, an **APPARENT ADDR IE** is enclosed in the **NEW** message sent by **S2** to **S3** and in the **ACCEPT** message sent by **S2** to **S1**. These information elements enclosed the Address of Contact of **S3** and **S1**.

10.6.4.4 Second Alternative

The second alternative is to delegate some functions to access IAX servers and not forward **NEW** messages to core IAX servers. **NEW** messages should be exchanged only between access IAX servers. Those servers communicate with the core IAX server, for instance to check the profile, verify access rules, update its call treatment and so on.

Figure 10.14 illustrates the IAX message exchange that occurs.

This proposal assumes the introduction of a new message, denoted **QUERY**. The goal of the introduced **QUERY** message is to retrieve the IAX server to which **B** is attached, and also to gain acceptance from **S2** to process the call. Some policies may be enforced in **S1** in order to avoid issuing this type of message systematically.

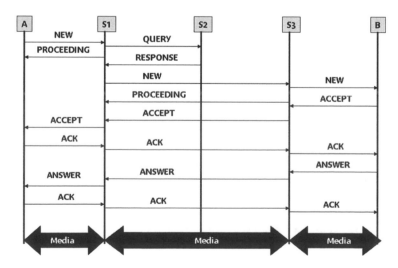

Figure 10.14   Path-decoupled call optimisation

This proposal is not part of the IAX specification but is an enhancement to the classical behaviour of the protocol. It aims to reduce the number of messages exchanged between IAX servers when a path-decoupled scheme is deployed within the administrative domain of a given VoIP service provider.

The use of the **QUERY** message is as follows:

- In order to place a call to **B**, **A** sends a **NEW** message.
- This message is received by **S1**, because it is **A**'s contact service point. A **PROCEDEEING** message is sent to **A** to notify it that the request is currently being handled by **S1**.
- Once the request is received, **S1** sends a **QUERY** message to **S2** to retrieve the appropriate information to deliver the service to **A**, and also to get the Address of Contact to which the **NEW** message should be forwarded to reach **B**.
- **S2** sends a **RESPONSE** to **S1**. This response includes the AoC of **S3** and additional policies.
- **S1** then forwards the **NEW** message to **S3**. **S3** sends the request to **B**.
- At this stage, the call involves **A**, **S1**, **S3** and **B**. The media path is therefore distinct from the signalling one which involves **S2**.

10.7 Forking

In some scenarios, and in order to optimise the required delay to place a given call, a given IAX server should be able to send several IAX requests to several adjacent IAX servers. This process may induce telephony routing loops. In order to implement this feature in the IAX environment, additional capabilities should be added to IAX so as to avoid these routing loops and implement a convergent routing process.

For illustration purposes, SIP supports this feature through exploitation of its **Via** headers. This avoids crossing the same server several times and drives the routing process.

10.8 Route Symmetry

SIP allows a symmetric telephony signalling path. By symmetric signalling path, we mean the capability of the signalling response path to follows the same route as the request. SIP implements this feature by making use of several headers, such as **Via**, **Route** and **Record-Route**. Within an interprovider context, only the interdomain path is symmetric; the internal path followed inside each crossed IP telephony domain will not be symmetric unless border nodes maintain states regarding ongoing sessions, and the path will then be followed by a received signalling response. This behaviour is not standardised and is not part of SIP itself.

Headers which implement route symmetry are not present in IAX; the routes experienced depend on underlying topology and configuration between servers. Nevertheless, IAX can be used jointly with appropriate resource-location protocols such as TRIP or DUNDi to implement a dynamic routing process.

10.9 Conclusion

IAX can be easily deployed in a path-decoupled scheme. But this mode is not a native process like SIP, since it requires acceptance from call participants in order to ensure that the call is successfully set up and that there are no connectivity issues. IAX is suitable for the access segments since VoIP service providers require the presence of service elements in the signalling and media paths (mainly to achieve some service-specific functions). Nevertheless, it is not recommend that IAX be used at the core segment within the same usage scenario as SIP; rather a suitable architecture that takes into account IAX specificities should be deployed. For instance, forking and route symmetry features are not supported by IAX. These features may be enforced jointly with other protocols such as TRIP.

References

[AES] US Department of Commerce/NIST, 'FIPS-197, Announcing the Advanced Encryption Standard', November 2001.

[DUNDI] Spencer, M., 'Distributed Universal Number Discovery (DUNDi)', draft-mspencer-dundi-01, October 2004.

[IAX] Spencer, M., Shumard, K., Capouch, B. and Guy, E., 'IAX2: Inter-Asterisk eXchange Version 2', draft-guy-iax-04, work in progress.

[RSA] Kaliski, B. and Staddon, J., 'PKCS #1: RSA Cryptography Specifications Version 2.0', RFC 2437, October 1998.

[SBC] Hautakorpi, J. et al., 'Requirements from SIP (Session Initiation Protocol) Session Border Control Deployments', draft-camarillo-sipping-sbc-funcs.

[SIP] Rosenberg, J., Schulzrinne, H., Camarillo, G., Johnston, A., Peterson, J., Sparks, R. et al., 'SIP: Session Initiation Protocol', RFC 3261, June 2002.

[TRIP] Rosenberg, J. et al., 'Telephony Routing over IP (TRIP)', RFC 3219, January 2002.

11

IAX and NAT Traversal

11.1 Introduction

NAT (Network Address Translator, [NAT]) has been adopted as the de facto standard for sharing a single IP address between several hosts. This technique was ignored for a while by standardisation bodies because it was believed to be against the spirit of the end-to-end argument [E2E]. Despite this, users have succeeded in imposing it on networks as a de facto standard as it meets a valid requirement (multiple devices reachable to one user). The IETF (Internet Engineering Task Force) tried to impose a new protocol (IPv6) [RFC1883] to solve these user requirements (plenty of available addresses), but it wasn't widely activated. In the meantime, NAT boxes are here.

The absence of a clear specification of NAT functions has led to a proliferation of implementations with 'fuzzy' behaviours. Each NAT implementation uses its own NAT algorithms and behaviours. As a consequence, several complications have arisen, and some patches have been proposed to bypass these intermediary nodes. The IETF recently launched a working group called BEHAVE (Behaviour Engineering for Hindrance Avoidance) to come up with a set of specifications for the behaviour of intermediary boxes such as NATs. BEHAVE's mission is to generate requirement documents and best current practices to enable NATs to function in as deterministic a fashion as possible. Before the creation of this working group, MIDCOM (Middlebox Communication) was chartered to specify protocols to ease middleboxes' traversal.

As stated above, the IETF has failed to 'shape' NAT function and promote interoperable and open implementation. Further, it has promoted protocols which suffer from rudimentary design flaws such as interference between OSI layers (for example TCP (Transport Control Protocol), which uses IP information to compute its checksum, and SIP (Session Initiation Protocol, [SIP]), which carries IP-related information). This interference, especially in the context of SIP, is a big problem when considering deployment scenarios. Indeed, several protocols, procedures and functions have been introduced to ease SIP NAT traversal. This chapter does not focus on SIP, since a large literature is available (e.g. [NATSIP]), but instead analyses issues raised when IAX [IAX] is used to access conversational services.

Inter-Asterisk Exchange (IAX): Deployment Scenarios in SIP-Enabled Networks   Mohamed Boucadair
© 2009 John Wiley & Sons, Ltd

11.2 Structure

This chapter is structured as follows:

- Section 11.3 describes several NAT types and provides a set of examples to illustrate the difference between them.
- Section 11.4 is dedicated to a generic discussion of issues related to IAX in the presence of NATs. Both client-hosted and server-hosted NAT perspectives are taken into account and described.
- Section 11.5 provides an overview of operational considerations. This section discusses NAT deployment issues, IP exhaustion problems and P2P concerns. An analysis of the applicability of IAX in all these scenarios is given.

11.3 NAT Types

11.3.1 Overview

In IP networking literature, plenty of NAT types are distinguished [STUN]. The main types are:

- *Full Cone (FC)*: for this type of NAT, all requests from the same internal IP address and port are mapped to the same external IP address and port. Any external host can send an IP packet to the host behind NAT using the mapped external address.
- *Restricted Cone (RC)*: for this type of NAT, all requests from the same internal IP address and port are mapped to the same external IP address and port. Unlike a full cone NAT, an external host (with IP address **IP1**) can send a packet to the internal host only if the internal host had previously sent a packet to **IP1**.
- *Port-Restricted Cone (PRC)*: a port-restricted cone NAT is similar to a restricted cone NAT, but the restriction includes port numbers. In other words, an external host can send a packet with source IP address **IP1** and source port **P** to the internal host only if the internal host has previously sent a packet to **IP1** and **P**.
- *Symmetric (SN)*: when a symmetric NAT is deployed, all requests from the same internal IP address and port to a specific destination IP address and port are mapped to the same external IP address and port. If the same host sends another packet with the same source address and port but to a different destination, a different mapping is used. In addition, only the external host that receives the packet can send a packet back to the internal host.

11.3.2 Examples of NAT Types

11.3.2.1 Architecture

In order to illustrate the difference between the NAT types given above, this section provides several examples. Figure 11.1 shows the architecture used to describe those examples. This architecture involves a set of telephones, each located behind a NAT box. Four types of NAT are considered. Furthermore, telephones with public addresses are also sketched in this figure.

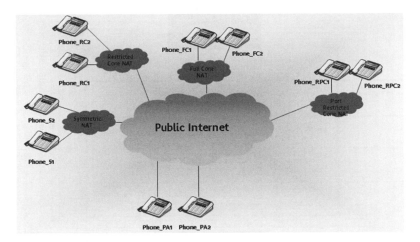

Figure 11.1 NAT scenarios

- **Phone_FC1** and **Phone_FC2** are behind a full cone NAT. The IP address of the public interface of this NAT is **212.25.26.25**. This address is used to translate the private IP addresses of **Phone_FC1** and **Phone_FC2**.
- **Phone_S1** and **Phone_S2** are behind a symmetric NAT. The IP address of the public interface of this NAT is **21.21.21.21**. This address is used to translate the private IP addresses of **Phone_S1** and **Phone_S2**.
- **Phone_RPC1** and **Phone_RPC2** are behind a port-restricted cone NAT. The IP address of the public interface of this NAT is **1.1.1.1**. This address is used to translate the private IP addresses of **Phone_RPC1** and **Phone_RPC2**.
- **Phone_RC1** and **Phone_RC2** are behind a restricted cone NAT. The IP address of the public interface of this NAT is **11.11.11.11**. This address is used to translate the private IP addresses of **Phone_RC1** and **Phone_RC2**.
- The public address of **Phone_PA1** is **35.26.25.25** and that of **Phone_PA2** is **15.25.35.45**.

11.3.2.2 Full Cone

When a full cone NAT is deployed, any remote machine can reach a node behind it if it knows a valid mapping already instantiated within it. To illustrate this behaviour, consider Figure 11.2.

In this example, **Phone_FC1** issues an IP packet destined to **Phone_PA1**. This packet crosses an FC NAT box, which instantiates an entry in its mapping table and modifies the source information, mainly the IP address and source port number, of the received packet. Thus, when exiting the FC NAT box, the IP packet is issued from a source IP address equal to **212.25.26.25** instead of **192.168.0.2**, with a source port number positioned to **1234** instead of the original value: **7855**. The packet is then delivered to **Phone_PA1** along an existing IP route.

As illustrated in Figure 11.2, **Phone_PA1** can use the perceived source IP information of received packets to contact **Phone_FC1** and therefore to send IP packets to **Phone_FC1**.

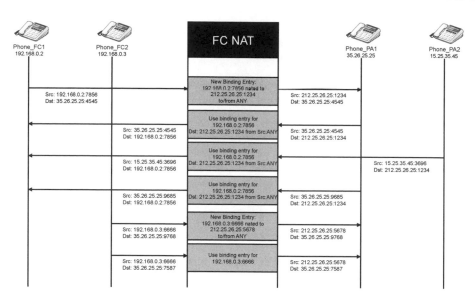

Figure 11.2   Example of full cone NAT behaviour.

These packets are received by the FC NAT box, which checks its mapping table and retrieves the original values of the destination IP address and port numbers. In this example, an entry is already instantiated; the FC NAT modifies the received packets and delivers them to their final destination. When exiting the FC NAT, the packets are destined to **192.168.0.2** instead of **212.25.26.25**, with a destination port number positioned to **7855**, according to the NAT mapping table.

When another machine, for example **Phone_PA2**, issues a packet destined to **212.25.26.25** with a destination port number equal to **1234**, this packet is received by the FC NAT box, which proceeds to the same operations as above. As a result, the packet is delivered to **Phone_FC1**.

This example shows that anyone on the Internet can issue traffic to an active NAT mapping, and that those packets will be delivered to machines behind a full cone NAT.

11.3.2.3 Restricted Cone

Unlike full cone NAT, restricted cone NAT accepts incoming packets only if a mapping entry to the source IP address exists in the NAT mapping table (that is, a packet has been issued before to that IP address).

To illustrate this behaviour, consider Figure 11.3. In this example, **Phone_RC1** issues an IP packet destined to **Phone_PA1**. This packet crosses a restricted NAT box, which instantiates an entry in its mapping table and modifies the source information, mainly the IP address and source port number, of the received packet. Thus, when exiting the restricted NAT box, the IP packet is issued from a source IP address equal to **11.11.11.11** instead of **192.168.0.2**, with a source port number positioned to **1234** instead of the original value: **7855**. The packet is delivered to **Phone_PA1** along an existing IP route.

As illustrated in Figure 11.3, **Phone_PA1** can use the perceived source IP information of received packets to contact **Phone_RC1** and therefore to send IP packets to **Phone_RC1**. These packets are received by the restricted NAT box, which checks its mapping table and

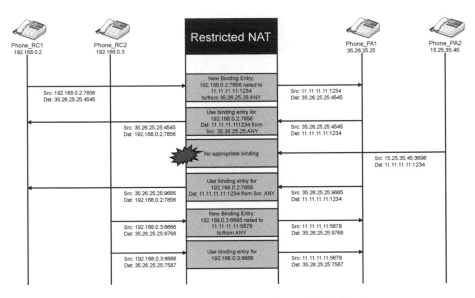

Figure 11.3 Example of restricted cone NAT behaviour

retrieves the original values of destination IP address and port number. In this example, since an entry is already instantiated, the restricted NAT modifies the received packets and delivers them packets to their final destination. When exiting restricted NAT, the packets are destined to **192.168.0.2** instead of **11.11.11.11**, with a destination port number positioned to **7855**, according to the NAT mapping table.

When another machine, for example **Phone_PA2**, issues a packet destined to **11.11.11.11** with destination port number equal to **1234**, this packet is received by the restricted NAT box, which proceeds to the same operations as above. But no entry is found in the NAT mapping table of the restricted NAT box so the packet is rejected. An ICMP (Internet Control Message Protocol) message may or may not be issued.

Note that if **Phone_PA1** issues a packet destined to **11.11.11.11**, with destination port number equal to **1234** and a source port number distinct from that used in the first issued packet of **Phone_RC1**, the packet will be delivered by the Restricted NAT to its final destination. No control is achieved based on the source port number.

11.3.2.4 Port-Restricted Cone

Unlike the restricted cone, this type of NAT accepts incoming IP packets only if the source IP address and the source port number have been used by an internal machine to send traffic (that is, only if both the port number and the IP address are indicated in the NAT mapping table).

Figure 11.4 illustrates the behaviour of port-restricted cone NAT. When a packet is received by the RPC NAT, it checks whether an entry is instantiated with both the IP address and port number. If no entry is present in the table, the packet is rejected.

Figure 11.4 shows two reject scenarios:

- **Phone_PA1** sends an IP packet to **Phone_RPC1** but uses a source port number distinct from the one **Phone_RPC1** used to send its initial packet (**9685** instead of **4545**).
- **Phone_PA2**'s IP address is not present in the mapping table.

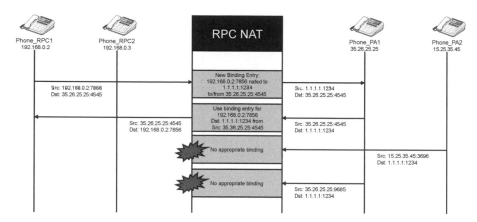

Figure 11.4   Example of port-restricted cone NAT behaviour

11.3.2.5 Symmetric

When symmetric NAT is deployed, all requests from the same internal IP address and port to a specific destination IP address and port are mapped to the same external IP address and port. If the same host sends a packet with the same source address and port, but to a different destination, a different mapping is used. Figure 11.5 provides an overview of the behaviour of a symmetric NAT box.

- When **Phone_S1** issues an IP packet with a source IP address **192.168.0.2** and source port number **7856** towards **35.26.25.25:4545** (respectively **15.25.35.45:8585**), the packet is handled by the symmetric NAT. This checks its mapping table and instantiates a new entry. It then translates the packet to a new one with source IP address equal to **21.21.21.21** and

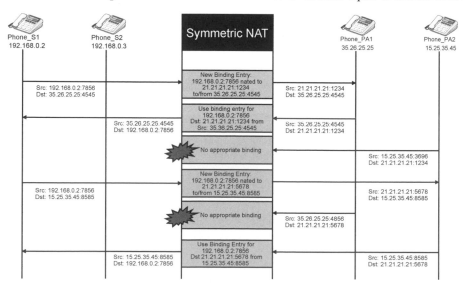

Figure 11.5   Example of symmetric NAT behaviour

source port number positioned to **1234** (respectively **5678**). The packet is then delivered to **Phone_PA1** (respectively **Phone_PA2**).

- **Phone_PA1** can contact **Phone_S1** using IP information contained in the received packet. When the NAT box receives the packet, it checks it mapping table. If a mapping entry is already present in the table, the packet is translated and delivered to **Phone_S1**.
- If **Phone_PA2** (respectively **Phone_PA2**) sends a packets to **21.21.21.21:1234** (respectively **21.21.21.21:5678**), the packet is rejected by the NAT function since no entry is instantiated for **15.25.35.45:3696** (respectively **35.26.25.25:5678**).

11.4 IAX and NAT Traversal Discussion

NAT is a complex technical issue to take into account when studying IP networks. The reader is invited to refer to [RFC3027] in order to become familiar with protocol complications induced by NAT boxes.

This section analyses the ability of IAX to bypass these complications and establish successful IAX sessions in various configurations.

11.4.1 Reference Architecture

In order to discuss NAT, several scenarios will be taken into account. Figure 11.6 shows three main ones:

1. Only the IAX server is behind a NAT box.
2. Only the IAX user agent (UA) is behind a NAT box.
3. Both the IAX UA and the IAX server are behind NAT boxes.

Obviously, study of the third scenario encompasses the first two cases. In the remainder of this chapter, only the third scenario is elaborated.

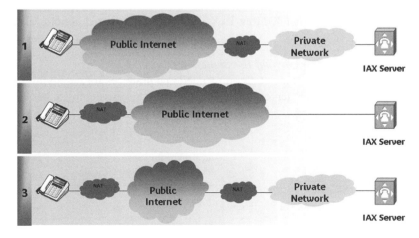

Figure 11.6   NAT presence scenarios

This scenario is generic even if it does not reflect current practices in terms of server deployment, which is usually done with public addresses or reachable addresses and without being located behind NAT boxes.

11.4.2 Discussion

In order to assess the ability of IAX to cross NAT middleboxes, the following section analyses, through several scenarios, registration and call establishment when NAT boxes are present in the path.

11.4.2.1 Signalling and Media Flows

The IAX protocol uses the same port number for both signalling and media messages. In this chapter no distinction is made between signalling and media messages, since a single port number is used.

11.4.2.2 Registration Considerations

This subsection is dedicated to registration issues. NAT discussions are undertaken from the perspective of both registrar and registrant.

IAX Registrar Issues
Two scenarios are considered: the presence of a single IAX registrar behind a NAT box and the presence of several registrars behind the same NAT box.

Single IAX Registrar
In order for IAX **REGREQ** requests to be routed to the IAX registrar, the NAT box (on the IAX registrar side) should be configured to accept all **REGREQ** from any (service) IP address and port number. The delivery of IAX requests may not succeed if the aforementioned NAT is not full cone (see Section 11.3). A static rule should be configured on the NAT side so as to forward all received **REGREQ** requests to the internal IP address and port number of the IAX registrar server.

Consequently, it is recommended that IAX registrars be deployed behind full cone NAT boxes.

Several IAX Registrar Servers behind the Same NAT Box
This scenario (see Figure 11.7) is not recommended and practices for deploying IAX servers should follow those for HTTP servers. If several IAX servers are deployed behind the same

Figure 11.7   Several registrars behind the same NAT box

NAT box, it is recommended that the port forwarding rules be configured so as to unambiguously identify the requested IAX registrar server. If they are not, IAX requests may not be delivered to the appropriate IAX registrar server.

Registrant Issues
An IAX registrant can always issue a **REGREQ** message to a given IAX registrar server. Whatever the type of the NAT box behind the IAX registrant, a **REGACK** message will be delivered to this registrant, because the IAX registrar server uses the perceived IP information (that is, source IP information of the received **REGREQ** message) to contact it back. In addition, the port binding at the NAT boxes is maintained by the exchange of periodical registration request messages (issued from the IAX client and destined to the IAX registrar server).

If the IAX server is behind the NAT, a port-forwarding rule should be configured so that these messages are forwarded to the IAX server.

In order to maintain NAT bindings (in the customer-side NAT), it is recommended that registration records be refreshed periodically so as to maintain the NAT binding.

11.4.2.3 Call Setup

This subsection focuses on the call setup phase. Two scenarios are considered:

- The same node is used as both IAX registrar server and proxy server.
- Distinct physical nodes are used to implement the registrar server and the proxy server.

Single Physical Node Hosts IAX Registrar and Proxy Server Functions
As illustrated in Figure 11.8, the same node is used to host both IAX registrar and proxy server functions. In such a case, once the registration procedure has succeeded, the call setup will lead to a successful call leg between the IAX client and the IAX server.

Two further scenarios should be studied:

- **Phone_1** wants to place a call to a remote destination. A **NEW** message should be issued by **Phone_1**. This request crosses the (customer-side) NAT, which applies appropriate translation operations. A modified **NEW** message is then routed until it is received by the (server-side) NAT box. Since the destination port number and destination IP address are the same as those of the previous **REGREQ** message issued by **Phone_1** (since the same physical node hosts both registrar and proxy server functions), whatever its type, the (server-side) NAT box accepts the request, applies appropriate modifications and delivers the message to the proxy server. For future messages, no problems will be experienced as the required NAT entries are already instantiated and active.

Figure 11.8 Registrar and proxy server hosted in the same machine

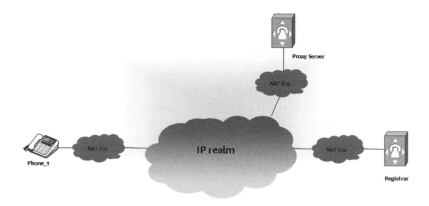

Figure 11.9   Registrar and proxy server behind distinct NAT boxes

- A **NEW** request to place a call with **Phone_1** is received by the proxy server. The proxy server relays this to **Phone_1**. The **NEW** message is received by the (server-side) NAT box, which proceeds to apply appropriate modifications and sends the modified **NEW** message to the public address of **Phone_1**. This message is then received by the (customer-side) NAT, which checks its mapping table and retrieves appropriate information to translate the received packet. A positive answer is retrieved from the mapping table, since **Phone_1** has already issued a **REGREQ** message to the source IP address and source port number of the **NEW** message. Once its destination information is modified, the **NEW** request is relayed to **Phone_1**. As for the response, it will be handled in the same way as described in the previous bullet.

IAX Registrar Does Not Handle Call Setup Request
Again, two scenarios should be considered:

- The IAX registrar server and proxy server are behind the same NAT box (see Figure 11.9). In this case, port-forwarding policies should be configured on the (server-side) NAT box so as to deliver received messages to the appropriate IAX server.
- The IAX registrar server and proxy server are behind distinct NAT boxes (see Figure 11.10). No port forwarding is required.

Both scenarios are similar to the first one discussed in Section 11.4.2.2. Thus, in order for IAX **NEW** requests to be routed to the IAX server, the NAT box (on the IAX server side)

Figure 11.10   Proxy server and registrar behind the same NAT box

should be configured to accept all **NEW** requests from any (service) IP address and port number. The delivery of IAX requests may not succeed if the server-side NAT is not full cone.

It is recommended that full cone NAT boxes be deployed at the server side so as to receive IAX messages from remote user agents.

Two further scenarios should be studied:

* **Phone_1** wants to place a call to a remote destination. A **NEW** message should be issued by **Phone_1**. This crosses the (customer-side) NAT, which applies appropriate translation operations. The modified **NEW** message is then routed until it is received by the (server-side) NAT box. Since the (server-side) NAT is full cone, the message is delivered to the IAX proxy server with no problems. For future messages, no problems will be experienced as the required NAT entries are already instantiated and active.
* A **NEW** request to place a call with **Phone_1** is received by the proxy server, which relays it to **Phone_1**. The **NEW** message is received by the (server-side) NAT box, which proceeds to apply appropriate modifications and sends the modified **NEW** message to the public address of **Phone_1**. This message is then received by the (customer-side) NAT, which should be full cone, or else at least one message should have been issued by **Phone_1** to the IAX proxy server.

11.5 Operational Considerations

11.5.1 Deployment Scenarios

11.5.1.1 Overview

In current deployments, service architectures are built around two main categories of nodes and functions (Figure 11.11):

* *Those located at the access segment*: responsible for managing service invocation, access control and so on. The first service node visible to customers is located at the access segment. All service-related flows are sent to this node.
* *Those located at the core segment*: responsible for the delivery of core service functions. These are generally hidden for end users and are not exposed to external parties.

These architectures may be based on current IMS and TISPAN specifications, or else be composed of a collection of servers following basic SIP specifications.

As far as IP connectivity is concerned, communication between service nodes may be isolated into a set of service VPNs (Virtual Private Networks), or it may be normally routed if a given service provider manages both service and IP infrastructure. Additional equipment, such as firewalls, is deployed to secure the service platform. IP addressing may be service-specific (an IP address is assigned when connecting to the service) or a single IP address may be used for all types of service.

With regards to NAT presence, communication between access nodes and core nodes is managed by a single administrative entity. No issues are experienced, even if no NAT boxes are deployed between these two service segments.

IP Telephony Domain

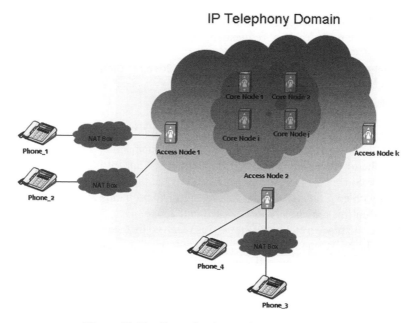

Figure 11.11   Example of operational deployment

From a customer perspective, NAT boxes may be used to connect to the service. The behaviour of these NATs is not standardised and may differ from implementation to implementation.

This section assumes that these customer NATs can be of any type. The following subsection provides an overview of the problems which occur when SIP or IAX is used to access the service.

11.5.1.2 SIP Use Case

The use of SIP to deliver conversational services has been investigated and solutions have been proposed. Indeed, several hurdles have been encountered, due to the path-decoupled nature of SIP and the interference between the service and the network layer. In particular, SIP carries information related to the network layer. For these reasons, several problems arise when crossing middleboxes, mainly NAT and firewalls. Plenty of solutions have been introduced, including:

- UPnP (Universal Plug and Play)
- STUN (Simple Traversal of UDP Through NATs, [STUN])
- Connection-Oriented Media
- Symmetric RTP
- TURN (Traversal Using Relay NATs)
- Media Relay (combination of Symmetric RTP and TURN server)
- ICE (Interactive Connectivity Establishment, [ICE])
- and so on.

In addition to these, which do not provide a global solution for all problems, dedicated equipment is introduced in the service platform to perform specific functions, such as hosted NAT traversal, performed by SBC (Session Border Controller, [SBC]) nodes. ALGs (Application Level Gateways) are also required at the home gateway node to modify the initial offer, including replacing the private address with the public one.

11.5.1.3 IAX Use Case

When using IAX instead of SIP in the delivery of conversational services (the context of Figure 11.11), no problems are encountered, since:

- A single IP address is provisioned in customers' devices to connect to the service.
- A single port number is used for both IAX control and media messages.

Whatever the type of the (customer-side) NAT, access to the service should be ensured without requiring additional patches or protocols.

11.5.2 The IP Exhaustion Problem

It is commonly agreed by the service provider community that the exhaustion of public IPv4 addresses is a fact. In this context, the community was mobilized to adopt a promising solution called IPv6 (Internet Protocol Version 6). Nevertheless, this solution is not globally activated by service providers, for both financial and strategic reasons. In the meantime, these service providers are not indifferent to the alarms recently emitted by the IETF (Internet Engineering Task Force). G. Huston introduced and promoted an extrapolation model to forecast the exhaustion date of IPv4 addresses managed by the IANA (Internet Assigned Numbers Authority). This model indicates that if the current tendency of consumption continues, the date of the exhaustion of IPv4 addresses in IANA's pool would be **2011**, whereas the one of that of those in the RIRs (Regional Internet Registry) would be **2012**.

In order to solve this exhaustion problem, service providers must investigate. In the meantime they should activate short-term solutions and continue to offer their IP-based service offerings. This section describes two solutions to the IP exhaustion problem, both of which optimise the usage of public IP addresses.

For more information about IP exhaustion, refer to Chapter 13.

11.5.2.1 Provider NAT

Solution Overview

Figure 11.12 provides an example of an architecture deployed by some service providers, mainly mobile service providers, to optimise the required public IP addresses for the delivery of added-value IP service offerings. This solution proposes introducing an additional level of NAT, denoted 'provider NAT' (also called 'double NAT'). This second level is hosted at the service provider perimeter.

This solution assumes that no public IP addresses are assigned to end users. Only private IP addresses are assigned to them (more precisely, to their equipment). When the traffic issued by

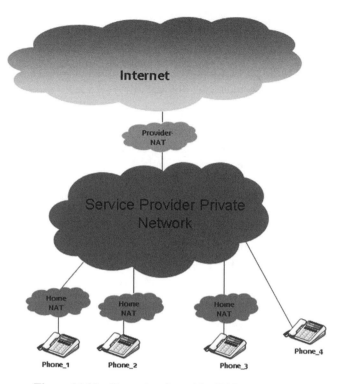

Figure 11.12   Example of provider NAT architecture

end-user terminals needs to exit the service provider private network, a NAT operation is required at the provider NAT box.

To illustrate this behaviour, consider Figure 11.13. When **Phone_1** wants to contact **Phone_4**, IP packets are issued with a source IP address equal to **10.0.0.1** and source port address **1234**. These packets are then transmitted to the home NAT, which proceeds to a NAT operation. When exiting the home NAT, the source IP address is positioned to **10.0.1.1** and source port number **5678**. These packets are then transferred across the private service provider network and routed to the provider NAT. This node proceeds to a second NAT operation. When exiting the provider NAT box, the source IP address becomes **25.25.25.25** and the source port number is **9123**. The packets are then routed to **Phone_4**.

In the rest of this chapter, it is assumed that the IP connectivity architecture is configured to maintain the IP transfer service with no alteration to current IP deployment.

SIP Use Case

In order to deliver SIP-based calls in the presence of provider NAT boxes, the following constraints should be followed:

• The service provider should be aware of the underlying IP infrastructure, so as to implement appropriate ALGs. At least two modifications to SIP messages should be applied: the first at the home NAT and the second at the provider NAT. If no such ALG is enabled, no

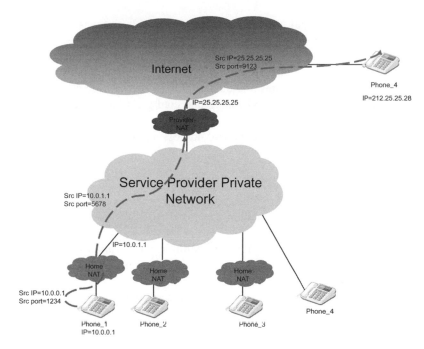

Figure 11.13   Example of provider NAT call flow

communication may be established. This constraint is 'heavy', since it assumes a vertical integration (that is, no functional separation between the service provider and the IP network provider) and that the same administrative entity administers both service and network infrastructure.

- NAT mapping entries at the provider NAT box should be maintained so as to deliver incoming message to customer devices located behind the provider NAT.
- Media flows may encounter some problems in delivery, since RTP ports may not be opened.

The introduction of provider NAT nodes may impact heavily on the delivery of SIP-based services.

IAX Use Case
When IAX is deployed, no ALG is required since no IP information is enclosed in IAX messages. Furthermore, the same port is used for both signalling and media messages.

If the provider NAT is configured to ensure global IP connectivity, no degradations are to be observed for IAX-based services.

The complexity is greater on the NAT provider side, to ensure a global connectivity with no IP-connectivity alterations.

11.5.2.2 Provider-Provisioned NAT

Solution Overview
As an alternative to the provider NAT solution, which suffers from several drawbacks, such as the alteration of already running services, the number of session states which need to be

maintained, the impact on legal aspects and so on, another option has been proposed. The motivation for introducing this second solution are as follows:

• To not alter current services delivery and not impact the introduction of future services.
• To avoid maintaining session states at the core network, privileging stateless solutions.
• To ease management (including provisioning and configuration operations) functions.
• To optimise CAPEX and OPEX.
• To make only a minor impact on routing and addressing architectures.
• To be transparent to end users.
• To ease usability.
• To facilitate functional separation (service and network).
• To ease implementation of legal requirements (optimise storage of legal data).
• To ease migration to a long-term solution such as IPv6.

This subsection focuses on the IPv4 variant of the solution. Other variants have been defined, to integrate IPv6 and offer a global connectivity service, including towards IPv6 realms, in a stateless manner.

The main idea of this second solution is to assign the same public address to several end-user devices and constrain the port number assigned by home NAT boxes. In addition to the IP address

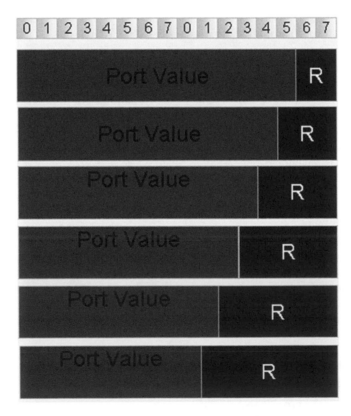

Figure 11.14   Examples of **Mask_Port**

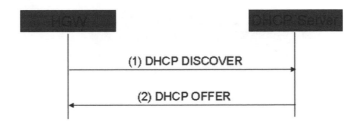

Figure 11.15   Example of DHCP exchange

assigned to access connectivity services, an additional parameter is conveyed in the session-initialisation message. This parameter is called **Mask_Port** (see Figure 11.14). This mask indicates which port range is to be used by customer devices. **Mask_Port** is coded in 16 bits.

If DHCP is used to convey configuration data to end users, a new DHCP option is defined to convey this **Mask_Port**, which is enclosed in the **DHCP OFFER** message (Figure 11.15).

Once received by a customer device (e.g. home gateway (HGW)), it constrains its NAT operations to the provisioned range. The number of customers to which an IP connectivity service provider can assign a single IP address depends on the number of allowed ports per user. If **N** bits are used to build the port mask (**R** bits in Figure 11.14), 2^N customers can be provided with the same IP address (for example, if N − − 3 the service provider multiplies its capacity, in terms of the number of customers to which the service can be delivered, by **8**).

A new function is introduced in the network, called 'Access Node'. This function takes charge of routing incoming flows to the correct home gateway when several share the same IP address. Therefore, IGP (Interior Gateway Protocol) should be tuned so that all incoming traffic must cross the Access Node. For routing purposes, a second IP address is assigned per home NAT. This is used to uniquely identify a given home NAT among all those with the same primary IP address. The secondary address is preferably a private one.

The Access Node is provided by (public IP address, **Mask_Port**, secondary IP address) so as to be able to proceed to port-driven routing. Figure 11.16 shows an example of IP communication when this procedure is adopted.

As shown in Figure 11.16, the same IP address, **25.25.25.25**, is assigned to the home NAT of **Phone_1** and **Phone_3**. A secondary address, **IP_Sec_1** and **IP_Sec_2**, is therefore assigned to each. Two port masks are also assigned to each user. In this example, these masks mean that that the home NAT of **Phone_1** can use a range of port number up to **10000**, and the home NAT of **Phone_2** can use a range of port number from **10100** to **20000**.

When **Phone_1** issues an IP packet to **Phone_4**, the source IP address is equal to **10.0.0.1** and the source port number is **1234**. Once the home NAT receives this, it proceeds to its NAT operations and assigns a port number in its provisioned range. In this example, a source port number **9123** is assigned. The packet is then routed to its final destination (**Phone_4**). **Phone_4** can send traffic to **25.25.25.25:9123**. This traffic crosses the Access Node, which proceeds to a port-driven routing. Concretely, the Access Node retrieves both the destination IP address and the destination port number from the received packet. Then it checks its binding table (list of (public IP address, **Mask_Port**, secondary IP address) entries) and retrieves the secondary IP address. The initial packet is encapsulated and sent to **IP_Sec_1**. Packets are routed to the home NAT of **Phone_1**, which proceeds to a de-encapsulation operation. At this stage it retrieves a packet destined to **25.25.25.25:9123**. As a final step,

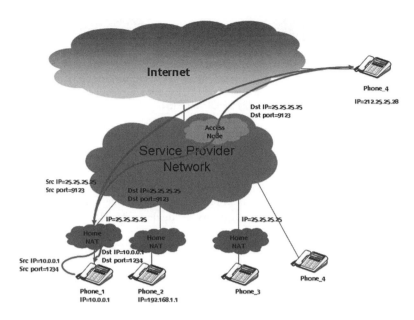

Figure 11.16 Provider-provisioned NAT

it checks its mapping table to find the local IP address and port number to be used. In this example, an entry is already instantiated; **10.0.0.1** and **1234** are returned and the packet is translated and routed to **Phone_1**.

All these operations are similar to classical NAT operations, except for the operations undertaken by the Access Node and the conditioned-port-numbers assignment process.

SIP Use Case

SIP-based services are not altered from current practices. The same mechanisms found in today's deployment are still used. No additional constraints or impacts are introduced.

SIP-based services are not altered and complexity is not increased.

IAX Use Case

IAX-based services are not altered from current practices. IAX-based services are not altered and complexity is not increased.

11.5.3 IAX, NAT and P2P Considerations

NAT is one of the critical technical problems met by P2P-based systems. Several techniques have been proposed to allow NAT traversal. Examples include:

- *Using a relay*: this technique uses a relay and reproduces a client/server-like architecture within a P2P system. Two remote peers may establish a communication through an intermediary node to which they already have an active communication, as illustrated in Figure 11.17. This technique consumes all the server's processing power and available bandwidth. Nevertheless, it is the most reliable technique and works whatever the NAT type.

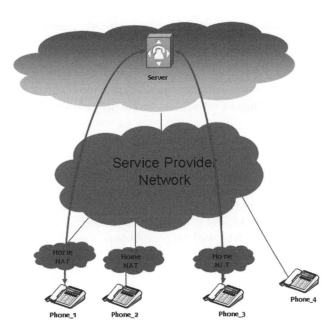

Figure 11.17   Example of NAT traversal using a relay

• *Connection reversal*: this technique uses an intermediary node to send connection-relayed
establishment requests. It is usually used when one of the peers involved is not behind a NAT.
Figure 11.18 shows the procedure when both peers are behind NAT boxes. Each initiates a
reverse connection request. These requests are relayed by an intermediary node. When the

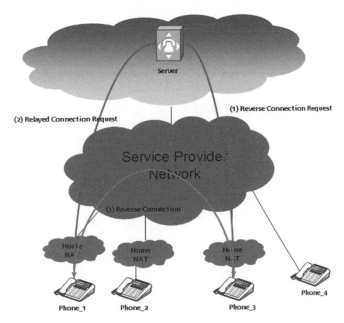

Figure 11.18   Example of NAT traversal using connection reversal

remote peers receives them, each attempts to make a reverse connection. As a consequence, a NAT entry is instantiated in each side, and a direct communication may be established between them.

If IAX is used to establish a session between remote peers in the context of P2P systems where one of the NAT traversal techniques is used, no additional problems are experienced.

11.6 Conclusion

This chapter discussed NAT traversal issues when the IAX protocol is activated for the delivery of added-value services, mainly conversational ones. The use of the IAX protocol does not introduce additional complexity to basic IP communication. This chapter also presented the IP exhaustion problem and two solutions to it. IAX can be activated in the context of these solutions, and it has been shown that it does not pose additional technical problems. Unlike SIP, IAX is powerful for NAT traversal and delivery of reliable communications.

References

[E2E] Saltzer, J.H., Reed, D.P. and Clark, D.D., 'End-To-End Arguments in System Design', ACM Transactions in Computer Systems, Vol. 2, Number 4, pp 277–288, November 1984.

[IAX] Spencer, M., Shumard, K., Capouch, B. and Guy, E., 'IAX2: Inter-Asterisk eXchange Version 2', draft-guy-iax-04, work in progress.

[ICE] Rosenberg, J., 'Interactive Connectivity Establishment (ICE): A Methodology for Network Address Translator (NAT) Traversal for Offer/Answer Protocols', draft-ietf-mmusic-ice-12, October 2006.

[IMS] Camarillo, G. and Garcia-Martin, M.A., *'The 3G IP Multimedia Subsystem: Merging the Internet and the Cellular Worlds'*, John Wiley and Sons, Ltd, 2005.

[NAT] Holdrege, M. and Srisuresh, M., 'Protocol Complications with the IP Network Address Translator', RFC 3027, January 2001.

[NATSIP] Schwartz, D. and Sterman, B., 'NAT Traversal in SIP', September 2005, available at: http://www.kayote.com/web/docs/Kayote_NAT_White_Paper.pdf.

[RFC1883] Deering, S. and Hinden, R., 'Internet Protocol, Version 6 (IPv6) Specification', RFC 1883, December 1995.

[RFC3027] Holdrege, M. and Srisuresh, M.,'Protocol Complications with the IP Network Address Translator', RFC 3027, January 2001.

[SBC] Hautakorpi, J. et al., 'Requirements from SIP (Session Initiation Protocol) Session Border Control Deployments', draft-camarillo-sipping-sbc-funcs-05.

[SIP] Rosenberg, J., Schulzrinne, H., Camarillo, G., Johnston, A., Peterson, J., Sparks, R. et al., 'SIP: Session Initiation Protocol', RFC 3261, June 2002.

[STUN] Rosenberg, J., Weinberger, J., Huitema, C. and Mahy, R., 'STUN: Simple Traversal of User Datagram Protocol (UDP) Through Network Address Translators (NATs)', RFC 3489, March 2003.

[TURN] Rosenberg, J. et al., 'Traversal Using Relay NAT (TURN)', work in progress.

Further Reading

Behavior Engineering for Hindrance Avoidance Working Group, http://www.ietf.org/html.charters/behave-charter.html.

Hain, T., 'Architectural Implications of NAT', RFC 2993, November 2000.

Middlebox Communication Working Group, http://www.ietf.org/html.charters/midcom-charter.html.

Raz, D., Schoenwaelder, J. and Sugla, B., 'An SNMP Application Level Gateway for Payload Address Translation', RFC 2962, October 2000.

Srisuresh, P. and Egevang, K., 'Traditional IP Network Address Translator (Traditional NAT)', RFC 3022, January 2001.

Srisuresh, P. and Holdrege, M., 'IP Network Address Translator (NAT) Terminology and Considerations', RFC 2663, August 1999.

Srisuresh, P., 'Security Model with Tunnel-Mode IPsec for NAT Domains', RFC 2709, October 1999.

Srisuresh, P., Tsirtsis, G., Akkiraju, P. and Heffernan, A., 'DNS Extensions to Network Address Translators (DNS_ALG)', RFC 2694, September 1999.

12

IAX and Peer-to-Peer Deployment Scenarios

12.1 Introduction

Peer-to-peer (P2P) services have recently emerged for the delivery of various types of service. P2P services do not rely on heavy centralised nodes. The service logic is distributed among several nodes. The main advantages of these architectures, compared to a centralised architecture, are the optimisation of CAPEX (Capital Expenditure) and OPEX (Operational Expenditure), and the enhancement of robustness and availability of service offerings. These new techniques have not been standardised and are mostly closed and proprietary solutions. This lack of standardisation is a hurdle for interoperability and interconnection. Furthermore, P2P service platforms are monolithic, and no mutualisation can be enforced without providing an interface to plug services over an overlay infrastructure.

Recently, possibly too late, standardisation fora have been interested in P2P-related topics, especially the IETF (Internet Engineering Task Force), for the delivery of SIP-based services, and 3GPP (3rd Generation Partnership Project), for the delivery of IP TV service offerings. As a result, a working group has been chartered within the IETF – the P2PSIP Working Group (WG) – to investigate P2P use in an SIP-based environment. This working group has encountered several problems and little progress has been made. This is mainly due to the need to focus the P2PSIP mission and clarify the relationship between SIP and the objectives of this WG as chartered by the IESG (Internet Engineering Steering Group). In the meantime, plenty of Internet drafts have been submitted to P2PSIP, but at the time of writing only two documents have been adopted as P2PSIP Working Group documents: the concepts document [CONCEPTS] and a protocol for locating resources [RELOAD].

- [CONCEPTS] defines concepts and terminology for the use of the Session Initiation Protocol in a P2P environment, where the traditional proxy servers and registrar servers are replaced by a distributed mechanism implemented using a DHT (Distributed Hash Table) or another distributed data mechanism with similar external properties.

Inter-Asterisk Exchange (IAX): Deployment Scenarios in SIP-Enabled Networks   Mohamed Boucadair
© 2009 John Wiley & Sons, Ltd

- [RELOAD] defines a P2P signalling protocol, denoted RELOAD (for REsource LOcation And Discovery). RELOAD is designed to provide a generic, self-organising overlay network service, allowing nodes to route messages to other nodes and to store and retrieve data in the overlay. Even though this protocol is defined within P2PSIP WG, it may be used with any signalling protocol requiring a location service. In particular, IAX can be used with RELOAD. For NAT (Network Address Translator) traversal issues, UDP (User Datagram Protocol) relays may be used. IAX transfer procedure should then be used to optimise the media path when required. NAT traversal within IAX architectures is further discussed in Chapter 11.

Unlike the P2PSIP initiative, this chapter focuses on the applicability of IAX to providing distributed and P2P conversational services for corporate customers. This is motivated by the need to activate open protocols to deliver highly-available and flexible services. In this context, IAX is used as the main signalling protocol. New objects and messages are introduced and defined to advertise and retrieve the locations of remote destinations, without requiring either registrar servers or proxy servers. Unlike Skype, which is forbidden in some administrations and enterprises for security reasons, the proposed solution in this chapter does not introduce new security issues. For the network administrator, firewall rules are simple to manage. More details about this solution are provided in the following sections.

12.2 Scope

This chapter focuses on an alternative solution for the implementation of P2P IAX services that is suitable for corporate customers. This solution does not require heavy DHT infrastructure and is based on native IP techniques in order to provide flexible and lightweight distributed conversational services.

This chapter also introduces a novel solution for the delivery of enriched telephony service offerings to corporate customers. This solution is based on multicast as the main connectivity mode for the deployment of a distributed and lightweight location service: MEVA (Multicast-Based VoIP Service Architecture). The MEVA architecture encloses both generic functions, to build and maintain the location service and associated routing primitives, and an open interface, to allow any signalling protocol to be plugged in above the MEVA location service, particularly the IAX protocol.

The purpose of the MEVA solution is to create a flexible IP telephony IP architecture, which is autoconfigurable and able to detect failures, and even to dynamically ensure service 'continuity' during those failures. This service offering targets organisations with no particular skills in managing telephony services. The MEVA service requires only the activation of multicast. Reliability and robustness functions are enclosed in the MEVA framework. For those organisations with no particular requirements on the control of the nodes which will act as 'Point-de-RendezVous' (PRVs), the dynamic mode can be adopted.

The intent of this chapter is not to provide a detailed specification of MEVA but rather to describe it in broad strokes, and to assess the applicability of IAX in this context.

12.3 A P2P Solution for Corporate Customers

12.3.1 MEVA Architecture

Figure 12.1 illustrates the reference architecture used within this chapter to illustrate the behaviour of MEVA. The underlying IP infrastructure is not detailed but it is assumed to be multicast-enabled.

MEVA does not require the deployment of additional equipment to implement and offer conversational services. Involved user agents (UAs) communicate between themselves to discover the location of remote UAs and to access offered services. DNS and DHCP [RFC2131] servers are deployed to avoid the provisioning of static information and to minimise configuration operation. Alternative methods, such as SLP (Service Location Protocol, [SLP]), can be also activated so as to discover the location of the MEVA service.

As mentioned above, the communication mode used to implement MEVA is IP multicast. As a reminder, this mode consists of sending an IP packet from a given source to a group of receivers. Owing to the multicast mode, the transfer of this IP packet is optimised, since the source issues only one packet even if several receivers are involved in the communication. That packet is then conveyed by the routers, which maintain a tree, called the 'multicast tree', which duplicates the packet when required, according to the distribution of group members. Multicast mode was designed to optimise bandwidth usage.

In order to build and maintain a multicast tree, specific routing protocols must be activated and configured in the underlying IP network. Some of these protocols, such as IGMP (Internet Group Management Protocol), are used to subscribe to a given multicast group, and others, such as PIM (Protocol-Independent Multicast), PIM-SSM (Protocol-Independent Multicast–Source-Specific Multicast), M-OSPF (Multicast Open Shortest Path First) and so on, are used to build the multicast tree itself.

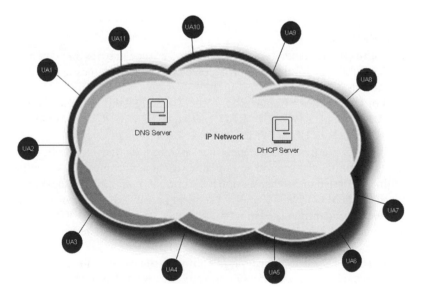

Figure 12.1 Reference architecture

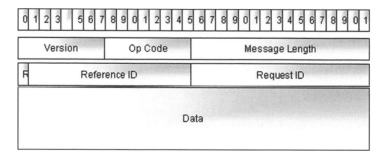

Figure 12.2   MEVA frame

12.3.2 MEVA-Related IAX Frames

Each MEVA message consists of the MEVA generic header, followed by a set of information elements depending on the nature of the MEVA operation. The structure of a MEVA frame is sketched in Figure 12.2.

The following fields must be enclosed when sending a MEVA frame:

- **Version** (8 bits): indicates the MEVA version used. The current version is **1**.
- **Op Code** (8 bits): indicates the type of the MEVA message. The following types may currently be used:
  1 = **ADVERT**
  2 = **ADVERT_REP**
  3 = **LEAVE**
  4 = **LOOKUP_REQ**
  5 = **LOOKUP_REP**
  6 = **MEVA_ERROR**
- **Message Length** (16 bits): indicates the size of the message in bytes, including the standard MEVA header and all encapsulated information elements. MEVA messages must be aligned on 4-byte intervals.
- **R** (1 bit): set to **1** for a request frame or **0** for other frames. So this field must be set to **1** when issuing **ADVERT** and **LOOKUP_REQ** messages and to **0** when sending **LEAVE, ADVERT_REP, LOOKUP_REP** and **MEVA_ERROR** messages.
- **Reference ID** (15 bits): stores an identifier assigned by the local MEVA speaker to unambiguously identify the session among all existent active ones. This identifier may be reused within the context of another session if the current session context has been destroyed. The length of this field is 15 bits. For all subsequent operations, this identifier must be used. Concretely, the same identifier used in **ADVERT** must be used in **ADVERT_REP, LEAVE** and related error messages. Furthermore, **ADVERT_REP** and related error messages must enclose the same identifier as the **LOOKUP_REQ**.
- **Request ID** (16 bit): used to convey an identifier assigned by a local MEVA speaker to identify a given request in the context of an active session (identified by a **Reference ID**). This field may be used when several requests have been sent to remote peers (e.g. forking). It must be set to **0** for **LEAVE**. The same request identifier used in **ADVERT** must be valued in

ADVERT_REP and related error messages. **ADVERT_REP** and related error messages must enclose the same identifier as the one indicated in the **LOOKUP_REQ**.
- **Data**: carries the data. The length of the enclosed data can be up to the maximum value supported by the network.

12.3.3 MEVA Information Elements

Within the context of MEVA, information element (IE) objects are used to carry useful information required for MEVA operations. These elements are carried in MEVA messages. For further discussion of information elements, refer to Chapter 5.

Table 12.1 groups a set of information elements which are useful within the context of MEVA. The following data is provided for each information element:

- *IE Name/Abbreviated Name*: indicates the full and the abbreviated name of the IE.
- *Description*: provides a brief description of the IE and its possible uses.
- *Related MEVA Messages*: lists the MEVA messages which may/must enclose this IE.

Table 12.1   List of MEVA information elements

IE Name Abbr. IE Name	Description	Related MEVA Messages
Called Number CALLED NUMBER	Within MEVA a called number is not limited to E.164 numbers but may also include non-numeric characters such as a SIP URI. The use of 'number' is misleading. This IE is used to carry the called URI (AoR)	LOOKUP_REQ
User Name USERNAME	This IE carries the identity of the user issuing a given MEVA message	ADVERT, ADVERT_REP, LEAVE, LOOKUP_REQ, LOOKUP_REP
Expire EXPIRE	This IE is used by a MEVA speaker to indicate the expire timer for a given event. The timer is expressed in seconds. The default value of this IE is 3600 seconds. The range of finite timeouts is 1 to 65 535 seconds, represented as an unsigned two-octet integer. The value of 0 implies infinity	ADVERT, ADVERT_REP
Call Number CALLNO	Within a transfer operation, this IE is used to indicate the number a remote peer needs to use as a destination call number to identify the ongoing call	ADVERT, ADVERT_REP
Cause Code CAUSECODE	This IE is used to indicate the reason for rejection of a request. Below are some cause	MEVA_ERROR

(*continued*)

Table 12.1   (*continued*)

IE Name Abbr. IE Name	Description	Related MEVA Messages
	code examples: 1 = Bad message format 2 = Incorrect identifier 3 = Unable to process 4 = Protocol error	
Called Context CALLED CONTEXT	This IE provides an indication of the remote dial plan context of the ongoing call. Note that a context may be line number, trunk group, etc.	ADERT, ADVERT_REP, LEAVE, LOOKUP
Service Identifier SERVICEIDENT	This IE is used to carry identifiers to uniquely identify services	ADVERT, ADVERT_REP, LOOKUP_REQ
Unknown UNKNOWN	When an unsupported MEVA method has been received by a MEVA peer, this IE must be issued	MEVA_ERROR
Apparent Address APPARENT ADDR	This IE is used by a given MEVA peer to indicate the apparent connection information (i.e. IP address and port number) of a remote MEVA peer	ADVERT, ADVERT_REP, LOOKUP_REP

12.3.4 How to Subscribe to a MEVA Service

This section details the steps which must be followed by any UA implementation in order to connect to the MEVA service:

• The first step is to provide an address IP to a given UA so it will be reachable within the enterprise LAN. This is achieved, for instance, through DHCP.
• The second step is to discover an IP multicast address, to be used to access the MEVA service. For this purpose, DNS is used to provide a resolution of the MEVA service FQDN. A dedicated message is then issued to connect to that multicast group.
• Once connected to the MEVA multicast group, the UA sends its publication message to subscribe to the MEVA service. This publication may be made either with or without acknowledgment. A dedicated message denoted as **ADVERT** is used to publish the contact information to remote UAs connected to the MEVA service. This message encloses the necessary information to contact a given UA in the context of a given session. **ADVERT_REP** is used to acknowledge this message. Numerous acknowledgement modes may be considered, but these are not described in this chapter.

12.3.5 Contact Table

Unlike traditional centralized architectures, the MEVA system is 'serverless'. In particular, no registrar or proxy servers are deployed to offer MEVA services. For this reason, a contact table is maintained by each UA participating in the MEVA service. Appropriate functions and operations should be implemented by these user agents.

The contact table (also denoted 'location table') stores all the contact information of remote UAs. It contains the information needed to establish a call with them. Its entries are instantiated or updated on the reception of an **ADVERT** or an **ADVERT_REP** message.

The following information must be stored in the contact table:

- *AoR (Address of Record)*: identifies the identity of a subscribed user. An AoR is encoded a URI.
- *AoC (Address of Contact)*: gives the IP address and port number to use to contact a user identified by a given AoR. This parameter may be enclosed in MEVA messages or retrieved from the headers of IP packets (especially source information).
- *Expire*: indicates the validity of an entry in the contact table.

It is also possible to add 'optional' fields to the contact table, such as the ones below:

- *Context*: indicates the context of the subscription to the service, such as 'business', 'friend' and so on.
- *Username*: indicates the username of the subscribed user.

12.3.6 Session Establishment

This section focuses on the call-establishment phase.

12.3.6.1 Retrieve the AoC of the Called UA

Once connected to the MEVA service, a UA can establish a call to a given destination. To do so, the calling UA has to know the AoC of the called UA. Two specific messages are introduced to retrieve the AoC of the called party.

LOOKUP_REQ()

This message is used to retrieve the AoC of the called party. This search operation is first done locally on the contact table of the caller. If the result is negative, it is sent to other UAs so they can retrieve the AoC of the called party. This message is issued using either multicast or unicast modes.

LOOKUP_REP()

This message is issued in response to a **LOOKUP_REQ** message. It encloses the AoC and additional appropriate information to contact the called party. Several responses may be received by the caller UA. Local policies are implemented to select one of them.

12.3.6.2 Call Establishment

Once the AoC of the called party is retrieved, the calling party issues a **NEW** message to that AoC. The same procedure described in Chapter 8 is followed to establish the call.

12.3.7 Leaving the MEVA Service

When a given UA wants to leave the service, a message **LEAVE(AoR)** is issued and sent to the MEVA multicast group. This message notifies remote UAs to update their contact tables. Upon receipt of this message, the remote UAs delete contact information related to that UA. Furthermore, in order to complete its service leave, an additional message denoted **LEAVEHOSTGROUP(@IP_MCTS)** is invoked and issued to notify the multicast router and update the multicast tree. Once issued, no MEVA-related traffic will be received by that UA.

12.3.8 Optimisation Issues

Several alternatives may be enforced to implement the AoR publication procedure within the MEVA service. This procedure strongly influences the call-establishment process, mainly to resolve the localization of a given UA within the MEVA service. In particular, if no acknowledgment messages are implemented, the problem of locating the AoC of a given destination UA is achieved by the originating UA itself, or by sending a multicast-based message to remote peers asking them to retrieve the targeted AoC. Within this scheme, all MEVA participants will receive a lookup message even if they are not involved in the call. All nodes will be flooded with lookup requests. This mode is not recommended. When an acknowledgment phase with a timer is enforced, and if the requirement is to avoid a multi-cast-based AoC search, the session establishment will involve the requesting UA. In this context, the lookup operation should be implemented in such a way as to lead to a convergent solution and avoid flooding the service nodes.

12.3.9 MEVA Architecture with a Point-de-Rendezvous

This section introduces a hierarchical mode which aims to structure the MEVA lookup operations and therefore optimise the required session setup delay.

12.3.9.1 Motivations and Needs

Several problems which can degrade the overall performances of the MEVA system may be encountered, such as overloading the service participants. Indeed, when the number of participating UAs becomes important, the number of maintained entries in the contact tables per UA and the number of treated requests will be 'huge'. In order to balance the overall traffic, reduce the number of exchanged messages and avoid multicast messages being flooded to every UA connecting to the service, a novel entity denoted 'Point-de-Rendezvous' (PRV) is introduced. The role of this entity is to centralise some of the functionalities hosted in the 'flat' model by all connected UAs. These newly-introduced entities should not be confused with the concept of a 'proxy server'.

The motivation for introducing this level of hierarchy is only to reduce the amount of exchanged traffic and not to reduce the complexity of the service logic, which remains similar in both modes. Moreover, this level of hierarchy is suitable for use in big organisations, where it can control the traffic issued by a given service and ease configuration of middleboxes such as firewalls and NATs (Network Address Translators).

The following subsections briefly describe some of the engineering issues for the MEVA service in the presence of PRV entities.

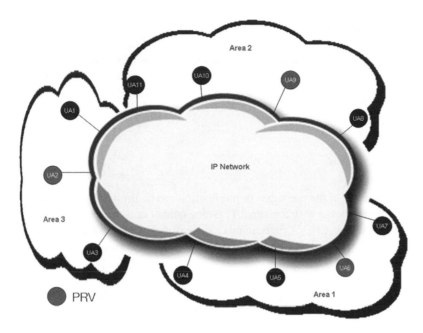

Figure 12.3 Example of MEVA service with areas

12.3.9.2 MEVA Areas and PRVs

What is a MEVA Area?

Overview

As described above, for optimisation purposes a new entity called 'PRV' is introduced. This entity is the central point for a set of UAs. Each set of UAs is located in the same zone, the scope of which is denoted 'MEVA area' or 'area'. The overall MEVA service is organised into several areas. Each of these is managed by a given PRV, as illustrated in Figure 12.3.

Figure 12.3 shows an example where the MEVA system is organised into three areas. For each of these areas a PRV is present (for example, the PRV of **Area 1** is **UA6**).

Several criteria may be used to define areas within a given MEVA system. For illustration purposes, two scenarios are described.

Geographic Areas

MEVA areas can be constituted in a geographical way. For example, all UAs located in a given city or country constitute a single MEVA area. From the perspective of an enterprise, a site located in a given geographic area can be used as a MEVA area. This structure logic is very interesting in the case of a company which has some subsidiaries in different cities or in various countries.

Functional Areas

Another possibility is to have functional areas 'mappable' with the functional organisation of a given company. Or areas could be organised per working team or per population, taking into account the frequency of telephony calls.

Engineering MEVA Areas
Overview
In order to implement MEVA areas, an important issue is how to organise the overall MEVA service so that connected UAs can be dispatched between several areas. Different options may be envisaged, including:

- TTL-based MEVA area.
- IP-based MEVA area.

TTL-Based MEVA Area
The reachability scope of the messages of publication can be limited by the field TTL (Time to Live). Indeed, an IP packet with a given TTL value cannot cross more routers than that value. Thanks to this technique, MEVA participants can be organized into several areas. The value of TTL must be positioned according to the underlying IP network topology.

IP-Based MEVA Area
Another method is to dedicate a specific IP multicast address to each area. By doing so, UAs must send their messages to that specific IP address. Only UA members of that area will receive those messages. This method is simple to administer and manage since no adherence with underlying IP topology is taken into account.

In order to avoid static provisioning, service IP multicast address per area may be advertised using SLP (Service Location Protocol, [SLP]), for instance.

Required Functions
Additional functions should be supported by the MEVA service in order to activate a dynamic mode for electing a PRV: essentially, the functions/procedures required to implement the PRV election process and its replacement when needed. More details about these functions are given below.

Electing a PRV
The PRV election function is one of the most important functions in a PRV-enabled MEVA system. It consists in assigning the PRV role to a given UA connected to the MEVA service. Several election logics may be envisaged. The challenge of this function is to let the service membership contribute in a deterministic way to the selection of the entity which will act as a PRV.

Notifying that a PRV Will Leave
When a UA acting as a PRV wants to leave the MEVA service, it has to notify all reachable peer UAs. A multicast message is issued to remote UAs to notify them of this event and to initiate a new procedure to select a new PRV. Two options may be envisaged to implement this feature:

- The first is that the UA acting as a PRV sends a dedicated message to notify peer UAs that it will leave the service and then disconnects immediately. For this variant, the service is unavailable until a new UA is selected to act as a PRV.

- The second is denoted the 'graceful mode'. The PRV sends a leave message to announce that it is going to disconnect after a given period **T**. During this period, it continues to act as a PRV while waiting for the MEVA service to select a new PRV. In parallel, an election procedure is initiated by the MEVA system. This option is advantageously recommended since service availability is ensured even during the replacement procedure.

Replacing a PRV
As mentioned above, when the UA acting as a PRV leaves the service 'brutally' or by sending a leave message, a process of replacement must be initiated. The objective of this function is essentially the election of a new and unique UA to act as a PRV.

Checking the Correctness of the Location Table
This function essentially helps to ensure the uniqueness of a PRV within the scope of a given area. If more than one entry in the location table with the field **is_PRV** is equal to **TRUE**, a procedure to select a single entity must be undertaken. This procedure depends on the election procedure.

Routing and Lookup Operations
Routing and lookup operations are executed according to two modes:

- *The PRV maintains a full contact table*: if this mode is activated, only local resolution is necessary to make a remote lookup response by the PRV. No lookup request is sent to remote PRVs. A lookup scenario is illustrated in Figure 12.4.
- *The PRV only has knowledge of its area and the PRVs of others areas*: in this mode, the lookup resolution can be extended to the PRV group if the initiator PRV does not store appropriate contact information regarding the called party (of course, a lookup request is to be sent in multicast only if the table of the local PRV indicates that adjacent PRVs are present in the system. If no adjacent PRVs are in its list then the request fails and a response is immediately sent to the requesting UA). Even in this scenario, two possible options may be implemented:

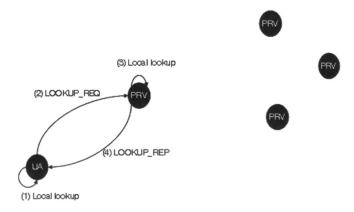

Figure 12.4 Full contact table mode

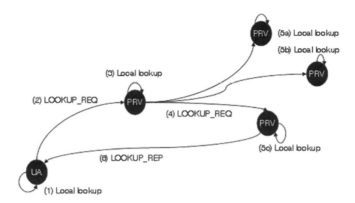

Figure 12.5   Multicast mode (direct-lookup-response mode)

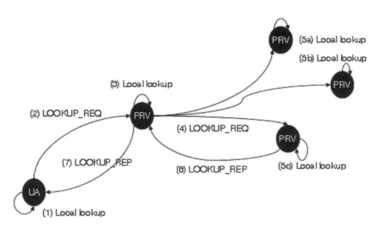

Figure 12.6   Multicast mode (to-originating-PRV mode)

- The remote PRV maintaining the contact entry of the called party sends a **LOOKUP_REP** directly to the initiator UA, as illustrated in Figure 12.5.
- The response is sent to the originating PRV, as illustrated in Figure 12.6.
- In order to implement the first mode, the lookup request issued by the originating PRV must include an **APPARENT_ADDR IE** enclosing the IP address and port number of the requesting UA.

Example of Call Setup
Figure 12.7 illustrates a call setup between **UA3** and **UA8** when a 'full table' mode is activated. Thus **UA3** first checks its local contact table. Because **UA8** is not in its local table, a **LOOKUP_REQ** is issued to its PRV (**UA2**). **UA2** checks its contact table and retrieves the AoC of **UA8**. This AoC is enclosed in a **LOOKUP_REP** message sent to **UA3**. Upon receipt of this message, **UA3** sends a **NEW** message to the AoC of **UA8** and the call can be established between **UA3** and **UA8**.

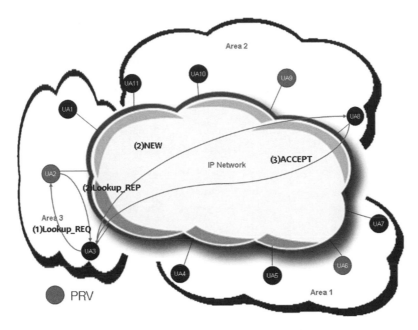

Figure 12.7   Example of MEVA call establishment

12.4 Conclusion

This chapter focused on P2P service offerings and the applicability of IAX to deliver P2P conversational services. A new architecture based on native IP capabilities has been introduced and defined. New IAX objects and messages have been defined to support distributed conversational services. This multicast-based P2P architecture does not require deployment of heavy DHT infrastructure.

The proposed architecture is suitable for implementation by corporate customers since it offer flexibility and simplifies required configuration operations.

References

[CONCEPTS] Bryan, D.,'Concepts and Terminology for Peer to Peer SIP', draft-ietf-p2psip-concepts, July 2008.
[RELOAD] Jennings, C.,'REsource LOcation And Discovery (RELOAD)', draft-ietf-p2psip-reload, July 2008.
[RFC2131] Droms, R.,'Dynamic Host Configuration Protocol', RFC 2131, March 1997.
[SLP] Guttman, E., Perkins, C., Veizades, J. and Day, M.,'Service Location Protocol, Version 2', RFC 2608, June 1999.

Further Reading

Ford, B., 'Peer-to-Peer Communication Across Network Address Translators', In USENIX Annual Technical Conference, 2005.
Peer-to-Peer Session Initiation Protocol Working Group, http://www.ietf.org/html.charters/p2psip-charter.html.
Schwartz, D., Sterman, B., 'NAT Traversal in SIP', September 2005, available at: http://www.kayote.com/web/docs/Kayote_NAT_White_Paper.pdf.
W. Fenner,'Internet Group Management Protocol (IGMP), Version 2', RFC 2236, November 1997.

13

IAX and IPv6

13.1 Introduction

The Internet community has been working for several years on migration issues related to IPv6 (Internet Protocol Version 6, [RFC1883]). A plethora of service providers envisage adopting IPv6 as the new connectivity protocol for many reasons, such as the abundance of addresses, its routing hierarchy spirit or the native auto-configuration features it supports. Furthermore, IPv6 has been adopted as the main IP protocol in several architectures, such as IMS (IP Multimedia Sub-System). Discussions regarding IPv6 activation and transition issues take place especially within the IETF (Internet Engineering Task Force), which hosts several working groups chartered to investigate operational and migration mechanisms.

This chapter does not argue in favour of introducing IPv6, nor does it examine how to benefit from the added value of this new protocol; it only discusses how a basic IP-based telephony service could be offered for both IPv4 [RFC760] and IPv6 customers. As far as IAX is concerned, this chapter analyses the support of IPv6 and the potential hurdles that may be encountered when deploying IAX in a heterogeneous environment, since this is part of the effort that should be undertaken by service providers to evaluate the compatibility of their service architectures with IPv6 and the impact of such a migration on their service offerings and associated architectures (for instance [RFC1933] and [RFC4038]).

This chapter focuses on telephony over IP services and the impact of the introduction of IPv6 on IAX-based architectures.

13.2 Context and Assumptions

The main goal of this chapter is to check if IAX-based telephony services could be offered to IPv6 customers, while ensuring coexistence with the already-deployed IPv4 IAX-based telephony service. It focuses only on intra-provider-related issues. Inter-provider VoIP issues are beyond the scope of this chapter. Authorisation-, authentication-, accounting- and end-to-middle-related issues are also out of the scope of this chapter. Indeed, it is not the purpose of this chapter to define new or improved architectures, but rather to examine from a generic point of view the ins and outs of such an interworking. End-to-end security and communication privacy are critical issues, but they too are beyond the scope of this chapter. The analysis of these issues could be inspired by the existing work related to NAT (Network Address Translation, [RFC3027]) and other middlebox solutions. Another important technical issue that is left for future investigation is the QoS (Quality of Service) offered by a given interworking mechanism.

Within this chapter, 'IAX service' denotes an IAX-based telephony service involving at least a proxy server and registrar server logical functions. In such a service, the proxy and the registrar servers have IPv4 and/or IPv6 connectivity. These functional entities could parse IPv4 and/or IPv6 addresses but could not use all of these IP versions as a transport protocol. A service provider offering an IAX service provides its customers with the information needed to reach IAX entities, especially the IP addresses of outbound and inbound proxy servers and the registrar server. These addresses could be real IP addresses or a FQDN (Fully Qualified Domain Name), or the address of an intermediary node that will isolate the service platform. This intermediate element would act as a relay to interconnect customers with the service-management platform. In this chapter, we do not make any assumption about the way a service provider configures the IP accessibility of its customers to the service platform.

This chapter does not recommend any particular IPv6 transition scenario, nor does it argue in favour of a particular IP transport interconnection mechanism to interconnect IPv4 and IPv6 realms. Nevertheless, it assumes that one of the possible mechanisms is activated, to allow communication between heterogeneous realms.

In the context of this chapter, the term 'heterogeneous' denotes two distinct IP version domains, a realm with IPv4 and a realm using IPv6. A communication between two nodes, each located in a distinct IP-version domain, is denoted 'heterogeneous' communication.

13.3 Service Migration to IPv6

In the context of activation of IPv6 at the transport layer, service providers should evaluate the impact of such a migration on their service-offerings portfolio and identify the required operations to avoid service alteration to their IPv6-enabled customers. This migration to IPv6 should be incremental and not implemented in one shot. For these reasons, service providers should come up with migration scenarios so as to achieve a transparent migration. This transparency is required because end users should not be aware of the underlying technology used to deliver their subscribed services, and complexity related to service engineering should be hidden from them.

As far as IAX is concerned, service providers should adopt clear strategies so as to ease the adoption of IPv6 and decrease the complexity related to IPv4–IPv6 interworking, which is one of the critical issues to be taken into account when designing service platforms. This chapter does not dwell on these transition considerations. Its main focus is on assessing the technical operations of an IAX service in a heterogonous environment.

13.4 Structure

This chapter is structured as follows:

- Section 13.5 provides a brief overview of the current IPv4 address allocations as undertaken by IANA. This section highlights the IP exhaustion problem encountered by the Internet community through a plethora of figures.

- Section 13.6 introduces IPv6 as the long-term solution to the IP address exhaustion phenomenon.
- Section 13.7 highlights the main technical problems met by telephony signalling protocols (especially SIP).
- Section 13.8 shows why IAX is an IP-agnostic protocol and how IAX does not carry information related to underlying layers.
- Section 13.9 describes how IAX can be deployed in a 'pure' IPv6 environment.
- Section 13.10 discusses the issues raised within a heterogeneous environment.

13.5 The IP Address Exhaustion Problem

It is commonly agreed by the service provider community that the exhaustion of public IPv4 addresses is a fact. In this context, the community was mobilised to adopt IPv6 as a promising solution to the problem. Nevertheless, this solution is not globally activated by service providers, for both financial and strategic reasons. In the meantime, these service providers are not indifferent to the alarms recently emitted by the IETF (Internet Engineering Task Force) (particularly by the reports presented within the GROW (Global Routing Operations) working group meeting (see the presentation of G. Huston, which predicts the exhaustion of available IPv4 addresses allocated by IANA at the end of 2009 [HUST]).

Figures [13.1–13.5] illustrate the state of the current 'consumption' of public IPv4 addresses. This information is updated daily and is available at www.potaroo.net/tools/ipv4/index.html.

Figures 13.1 shows the state of the IPv4 address allocations carried out by IANA for the RIRs (Regional Internet Registries).

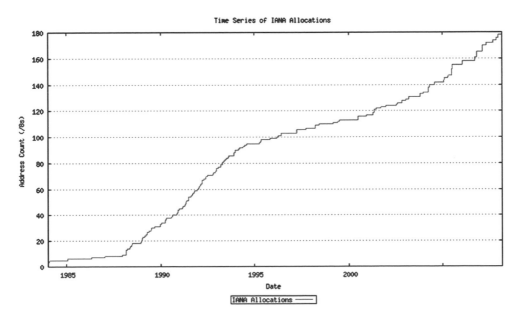

Figure 13.1 IPv4 IANA allocations (source: [HUST])

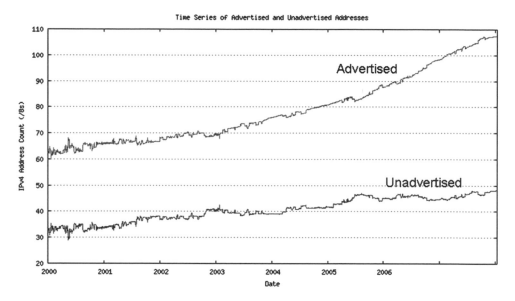

Figure 13.2   Advertised and unadvertised IPv4 addresses (source: [HUST])

In order to ease understanding of this graph, we specify that an IPv4 address is coded on **32 bits** (**4 294 967 296** possible addresses). This addressing space is divided into 256 **/8** prefixes, with each **/8** prefix corresponding to **16 777 216** single IPv4 addresses. Figure 13.1 shows that currently IANA allocations are close to 180 **/8** prefixes. Note that almost **32 /8s** are reserved for the IETF.

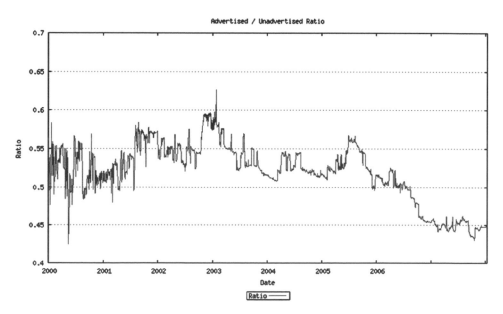

Figure 13.3   Advertised:unadvertised ratio (source: [HUST])

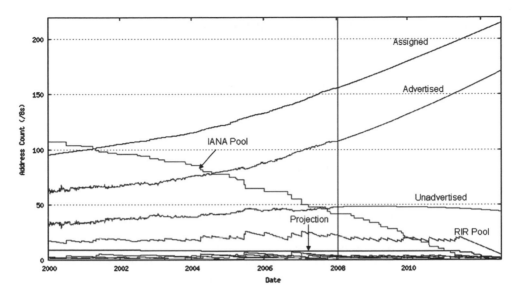

Figure 13.4   Address consumption model (source: [HUST])

The situation would not be so alarming if the allocated addresses were not announced on the Internet (that is, if there was effective use of the addresses). Figure 13.2 shows that the number of addresses announced in the routing protocols is increasing, and Figure 13.3 shows that the ratio is close to **0.5**. This means that 90–95% of IPv4-allocated addresses are visible in the routing tables.

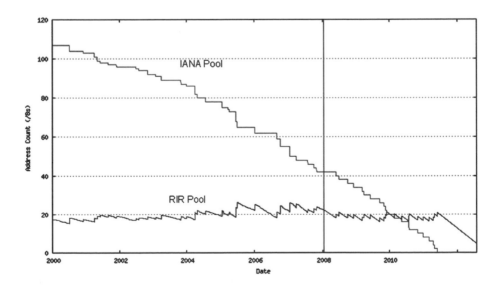

Figure 13.5   Projected RIR and IANA consumption (source: [HUST])

G. Huston introduced and promoted an extrapolation model to forecast the exhaustion date of IPv4 addresses managed by IANA. This effort indicates that if the current tendency of consumption continues as it is, the date of the exhaustion of IPv4 addresses in IANA's pool would be **2011**, whereas that of the RIRs would be **2012**. (see Figures 13.4 and 13.5).

It should be noted that the current model might even be 'catastrophic' if service providers and telcos, heeding these alarms, hurry to ask for more IPv4 prefixes to use in the event of an IPv4 shortage. In that case, the IPv4 address exhaustion date would be closer than that envisaged and forecasted by the aforementioned model.

In order to solve this depletion problem, service providers needs to investigate and enable methods to ensure the deployment of their service offerings and delivery to end users. Two options may be envisaged:

- Migrate to IPv6. This option requires deploying mechanisms for interconnection with the already-existing IPv4 realms.
- Enhance IPv4 and optimise the assignment of IPv4 addresses. An example of the implementation of this option is the introduction of a second level on NAT, called Provider NAT. With such an option, only private addresses are assigned to end-user home gateways. A Provider NAT node will be responsible for translating packets issued with private addresses into ones with publicly-routable IPv4 addresses. This node is located in the service provider domain. When deploying such a scenario, several hurdles will be encountered, for example:

  - End users won't be able to configure their own port-forwarding policies anymore.
  - It will be necessary to activate a second ALG (Application Level Gateway) at the core network.
  - There will be problems running servers behind domestic NATs.
  - There will be complications when maintaining NAT entries.
  - And so on.

More details regarding Provider NAT (also denoted 'Double NAT') are given in Chapter 11.

In order to solve the problem of IPv4 exhaustion, an alternative proposal has been made within the IETF to release IPv4 class **E** addresses [CLASSE]; concretely, to reclassify **240/4** as usable unicast address space. The rationale behind this document is that since the community has not concluded whether the **E** block should be considered public or private, and given the current consumption rate, it is clear that this block should not be left unused.

This proposal requires updating IP-enabled equipment so as to correctly treat IPv4 addresses belonging to **240/4** blocks. These addresses should be routable, and announced for instance between adjacent autonomous systems (ASs) through a BGP (Border Gateway Protocol). An exhaustive study should be undertaken to evaluate the economic and technical impact of such a new policy.

13.6 IPv6: a Long-Term Solution

13.6.1 Overview

IPv6 has been around for several years as the next version of protocol IP. This new version offers an abundance of IP addresses as well as several enhancements compared to IPv4, especially with the introduction of hierarchical routing (and therefore a reduction in the routing table size). IPv6 specifications are mature and current work within the IETF is related to operational aspects. Nevertheless, service providers have not yet activated IPv6 in their networks. Moreover, even if a service provider were to activate IPv6, it will be confronted with the problem of ensure total connectivity with the Internet v4 of today. Mechanisms such as NAT-PT (Network Address Translation Protocol Translation) and the 'Tunnel Broker' were introduced to ensure interconnection between two heterogeneous realms (that is, IPv4 and IPv6) and to ensure a continuity of IP communications (that is, end-to-end).

Despite the current IPv6 deployment situation, IPv6 is a long-term and viable alternative for offering IP connectivity services to a large number of customers. From this perspective, service providers should avoid introducing new functions and nodes which might be problematic when migrating to IPv6. This critical requirement should not only be taken into account during the technical engineering phase but also when elaborating the required CAPEX/OPEX estimation of the activation of alternative schemes to solve or reduce the impact of the IPv4 address exhaustion phenomenon.

13.6.2 Between IPv4 and IPv6: Where Is IPv5?

The Internet Stream Protocol (ST) is an experimental connection-oriented internetworking protocol that operates at the same layer as connectionless IP. Both the Internet Stream Protocol and IP apply the same addressing schemes to identify different hosts. Internet Stream Protocol and IP packets differ in the first four bits, which contain the protocol version number. Number 5 is reserved for Internet Stream Protocol: IPv5. The first version of the Internet Stream Protocol was published in the late 1970s, and a second version was specified in RFC1819, but neither has ever been deployed.

During the design of IP next-generation (which later became IPv6), number 5 was not assigned since it was attached to the Internet Stream Protocol.

13.6.3 IPv6 at a Glance

Below are some of the technical characteristics of IPv6:

- Unlike IPv4, IPv6 defines a new address scheme encoded in 128 bits. Therefore, the number of available IPv6 addresses is **340 282 366 920 938 463 463 374 607 431 768 211 456,** compared to **4 294 967 296** possible IPv4 addresses. This means there are **79 228 162 514 264 337 593 543 950 336** times more IPv6 addresses than IPv4 addresses.
- Unlike IPv4, IPv6 does not support broadcast. Indeed, IPv6 defines the notion of multicast scope.
- Unlike IPv4, only the sender of traffic may proceed to packet fragmentation.
- IPv6 has a simplified header. All 'optional' fields of IPv4 have been replaced by 'option headers'.
- IPv6 support native methods for autoconfiguration.

Contrary to some IPv6 'propaganda', IPv6 is not different to IPv4 in terms of QoS and security. Indeed, IPv6 is no more 'secure' than IPv4 and does not 'create bandwidth'! QoS-related fields are not the same in both headers but the mechanisms to enforce QoS may apply in both IP versions. IPSec suite is compatible with both IPv4 and IPv6. The problem is more related to the availability and deployment of a universal and global security infrastructure than the support of IPv4 or IPv6.

Finally, it is worth mentioning that migrating to IPv6 is an issue for service providers, not their customers. Customers (both residential and corporate) require global IP connectivity. How this connectivity is engineered and put into effect is the business of IP connectivity service providers. Of course, some 'corporate' customers will want to specify the nature of their IP connectivity and reduce the interconnection engineering complexity of their interconnection nodes within the domain of their IP connectivity service provider(s). From this standpoint, service providers should be more proactive in order to avoid a crisis scenario where no IP addresses are available to be assigned to their customers.

13.7 Why IPv6 May Be Problematic for Telephony Signalling Protocols: the SIP Example

13.7.1 Overview

SIP protocol is a signalling protocol that has been standardised by the IETF in order to initiate multimedia sessions. SIP is used for initiating, modifying and closing sessions. SIP works jointly with the SDP protocol (Session Description Protocol, [RFC2327]). The latter is used to describe multimedia parameters like the address and port number that will be activated for RTP streams. In addition, SIP manipulates IP addresses in order to route signalling messages. These addresses can be found for example in the **contact** header, the **request URI**, **Via** or in the SDP headers, especially **originator**, **contact** and **media description**.

Since its first version, documented in RFC 2547, IPv6 addresses have been supported by SIP. Thus, SIP implementations should be able to parse IPv6 addresses. These addresses can be used as AoR (Addresses of Record). From this perspective, the IPv6 protocol is not a problem in terms of parsing and encoding operations. Nevertheless, the issue is more related to preventing session-establishment failure when endpoints are not located in the same IP-version routing realm.

SIP is a signalling protocol used to set up multimedia sessions. Its main function is to allow user agents to successfully exchange media streams. Therefore, SIP manipulates 'pieces' of data that are meaningful for user agents. In particular, SIP messages carry information which is meant to be exploited by user agents to determine their interlocutors' supported media type and therefore converge on common media codecs and settings, allowing the establishment of multimedia sessions.

13.7.2 Additional SIP Tags including IP-Related Information

The purpose of this section is to highlight some issues related to the use of optional SIP tags which influence the routing behaviours of subsequent SIP messages. Some of these tags have been introduced in documents more recent than [SIP] in order to solve problems raised when deploying SIP services in different environments, for instance behind NATs. A comparison could be drawn between NAT handling and IPv4–IPv6 interworking as the base of both problems is to handle two different addressing schemes.

13.7.2.1 maddr

This tag can be present in **Via** headers. If it is present, responses must be sent to the address set in this tag, with the port number set to the value designated in the **sent-by** field of the same **Via** header, or **5060/5061** by default. This tag should be taken into account when investigating IPv4–IPv6 interworking solutions. Its impact should be carefully studied and assessed.

13.7.2.2 rport

This tag was introduced in [RFC3581] to solve the problem of NAT traversal. When SIP is used over UDP (User Datagram Protocol), the responses to SIP requests are sent to the IP address from which the request came and the port number which is written in the topmost **Via** header of the request.

SIP has defined methods to detect whether the IP address value in the **Via** header is the same as that the request came from, by using the **received** tag. Therefore, a given UA checks whether the source IP address of the received IP packet is the same as the value written in the topmost **Via** header. If these two addresses are distinct, the UA sets the value of the **received** tag to the source IP address of the received SIP message. The value of the **received** tag will be used for sending responses to these requests. This behaviour is useful for NAT traversal. Nevertheless, in [SIP] there is no similar mechanism for checking if the source port is the same as the one written in the topmost **Via** header. This could lead to problems where responses are sent to the wrong port values. In order to handle this issue, [RFC3581] suggests an analogous parameter to **received,** containing the source port number.

13.7.2.3 anat

[RFC4091] describes a mechanism to allow multiple alternative network addresses to be enclosed in a single SDP offer. This proposal consists of introducing a new attribute called 'Alternative Network Address Types' (ANAT). This attribute allows insertion of multiple media lines in the same SDP message.

[RFC4092] defines how SIP can exploit the ANAT semantic by introducing a new option tag called **sdp-anat**. This tag can be used by user agents to discover each other's capabilities and then select from the supported media description lines the ones that are suitable for setting up the SIP communication. One use case for this tag is a dual-stack user agent which can communicate either in IPv6 or IPv4. This user agent can also set a preference associated with each type of media. The proposal described in [RFC4092] gives guidelines to ensure correct processing of this new tag by user agents not supporting the ANAT semantic.

13.7.3 IPv6-Embedded SIP Examples

In order to illustrate the results of SIP session-establishment procedures between heterogeneous user agents, this section provides a set of examples of SIP call flows when only one IPv4-enabled proxy server and IPv4-enabled registrar are deployed, denoted respectively by **PS** and **R**. These elements have only IPv4 connectivity.

We suppose that the IPv6-enabled UAs have been provided with the relevant information to contact both **PS** and **R**. For the purpose of this example, problems related to IPv6 UA registration to **R4** and exchanges with **PS4** are assumed to have been handled.

Figure 13.6   Registration call flow

13.7.3.1 Registration

Figure 13.6 lists the SIP messages exchanged when **B** registers in **R**. In this example, we assume that **R** can parse and store an AoR including an IPv6 address.

Here are the details of the content of these messages:

1. **B** sends a **REGISTER** message to **R** (Table 13.1).
2. **R** sends back an **OK** message to confirm the registration (Table 13.2).

Table 13.1   Example of a **REGISTER** message.

```
REGISTER sip:r.test.com SIP/2.0
Via: SIP/2.0/UDP [2001:688:1ffb:ff80::2]:5062;branch=z9hG4bK00e31d6ed
Max-Forwards: 70
Content-Length: 0
To: Test <sip:test@test.com>
From: Test <sip: test@test.com>;tag=ed3833bd7363e68
Call-ID: a8a83b610ae5d242289dfc1c78b7f1d8@test.com
CSeq: 1830746364 REGISTER
Contact: Test <sip:test@[2001:688:1ffb:ff80::2]:5062>;expires=900
```

Table 13.2   Content of **OK** message

```
SIP/2.0 200 OK
Call-ID: a8a83b610ae5d242289dfc1c78b7f1d8@test.com
CSeq: 1830746365 REGISTER
From: Test <sip: sip:test@test.com>;tag=ed3833bd7363e68
To: Test <sip: sip:test@test.com>;tag=3ab7fe89d998709
Via:SIP/2.0/UDP[2001:688:1ffb:ff80::2]:5062;received=10.16.15.29;
branch=z9hG4bK00e31d6ed
Content-Length: 0
Contact: Test <sip:test@[2001:688:1ffb:ff80::2]:5062>;expires=900
```

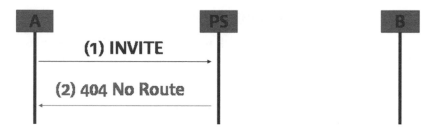

Figure 13.7   Example of a call flow

Once this registration process is achieved, **B** is reachable at the address indicated in the contact, which is stored together with a port number in the registration database.

13.7.3.2 An IPv4-Enabled UA Calls an IPv6-Enabled UA

The exchange of SIP messages that will occur when an IPv4-enabled UA **A** tries to initiate a SIP session with the IPv6-enabled UA is described in Figure 13.7.

When **A** sends an **INVITE** message to its **PS**, the latter requests its routing process to find a route towards **B**'s AoR. But since the AoR of **B** is an IPv6 address, **PS** isn't able to forward the request to **B**. An error message is then sent back to **A**. The type of the error message depends on the implementations. As a consequence, it is impossible to initiate a call destined to an IPv6-enabled user agent.

13.7.3.3 An IPv6-Enabled UA Calls an IPv4-Enabled UA

Let's suppose now that the IPv6-enabled user agent (**B**) sends an **INVITE** message to an IPv4-enabled user agent (**A**). Figure 13.8 illustrates the beginning of the SIP message exchange that occurs.

The behaviour of **A** depends on the UA's implementation. When receiving the **INVITE** message, **A** parses the SDP part and formulates its answers. This could lead to a closure of the application or simply to sending back an error message because the media line of the SDP part doesn't include an IPv4 address. As a consequence, it is impossible to set up an SIP session initiated by an IPv6-enabled UA towards an IPv4-enabled UA.

More discussions related to IPv6 complications and proposed solutions related to SIP are given in [BOUCA08].

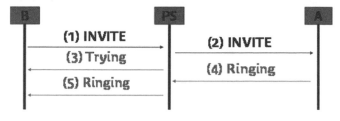

Figure 13.8   Example of a call flow

13.8 IAX: an IP Version-Agnostic Protocol?

As stated above, SIP encloses in its messages IP-related information. This interference between the service layer and the network ones introduces additional complexity in the placement of calls between heterogeneous user agents.

Unlike SIP, most IAX messages do not carry any IP addresses nor information about the port number. IP-related information may optionally be enclosed in some response messages, such as **REGREQ** or **REGACK,** to inform the remote peer about the perceived IP-related information (that is. source IP address and port number), or in **TXREQ** to indicate the apparent address to be used during the tentative transfer. IP-related information is indicated as part of a **TXREQ** message so as to help remote participants 'open' NAT entries on both sides and then ease the establishment of direct communication between these two participants. Note that these addresses may be IPv4 or IPv6.

From this standpoint, IAX is an IP-agnostic protocol and no interference is experienced between the service and the network layer.

If appropriate mechanisms have been put in place to interconnect IPv4 realms with IPv6 ones, and as far as IAX adopts a path-coupled scheme, IAX message exchanges can lead to successful call sessions. Only some optional features such as call optimisation may fail. A detailed discussion is elaborated below.

13.9 Deployment of IAX Services in a 'Pure' IPv6 Environment

In this section, we focus on a full IPv6 scenario, as illustrated in Figure 13.9. This scenario is similar to a full IPv4 one. All IAX operations (registration, call setup, call optimisation and so on) remain the same as per IPv4. Within this scenario, no issue is raised and the current specifications of IAX are coherent with a pure IPv6 environment.

In this scenario, the deployment of an IAX service over a 'full' IPv6 infrastructure is assumed. In this context, both endpoints (that is, user agents) and IAX service contact addresses are identified by their IPv6. The underlying IP infrastructure is able to deliver packets from (to) an IPv6-enabled user agent to (from) an IAX server.

The IAX URI of **A** may enclose an IPv6-address, as shown in Table 13.3.

Furthermore, IAX control messages such as **REGREQ**, **REGACK** and **TXRQ** may include an IPv6 address in a dedicated information element called **APPARENT ADDR IE.**

In order to avoid parsing errors, the IPv6 address is enclosed between '[' and ']', so as to delimit it from the port number and not to confuse the ':' used with the one which separates an IP address from a port number in an IAX URI.

In Figure 13.9, **A** is provided with either an IPv6 address of **S1** or an FQDN. If n FQDN is provided to reach the IAX service, **A** should issue a **AAAA** DNS query to retrieve the IPv6

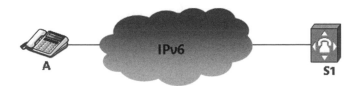

Figure 13.9   IAX in a full IPv6 environment

Table 13.3   IAX URI of A

```
iax:[2001:688::1]:4569/A?work
```

address of **S1**. Once retrieved, **A** sends its regular IAX message to **S1** using IPv6. The underlying IPv6 infrastructure takes charge of the delivery of issued IPv6 messages to their final destinations. Consequently, **A** may register to an IAX service and place/receive calls forwarded by **S1**.

This leads to the conclusion that an IPv6-enabled service is not altered and that the same level of service as per IPv4 will be experienced by a given customer within a 'full' IPv6 environment.

13.10   Heterogeneous Environment

13.10.1   Context and Reference Architecture

The goal of this section is to study the impact of the coexistence of IPv4 and IPv6 in the core IP network on IAX-based communications.

Figure 13.10 illustrates two simple configurations in which IPv4 and IPv6 realms are interconnected through an 'interconnection node' (IN). This IN may be a physical node or simply one of the transition mechanisms specified within the IETF. For more information about these transition mechanisms, the reader is invited to refer, for instance, to [RFC1933] and [RFC4038]. We assume within this book that one of the IPv6 transition mechanisms is enabled to interconnect the IPv4 realm with an IPv6 one.

Because the two scenarios described in Figure 13.10 are similar, only the cases where the IAX server is attached to an IPv6 realm and where that server is dual-stack are detailed and elaborated. These two scenarios are illustrated in Figure 13.11.

The entities involved are:

- **A:** an IAX user agent attached to the IPv4 realm.
- **B:** a dual-stack IAX user agent
- **S2:** a dual-stack IAX server.

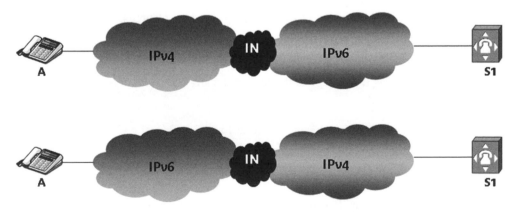

Figure 13.10   IAX in a heterogeneous environment

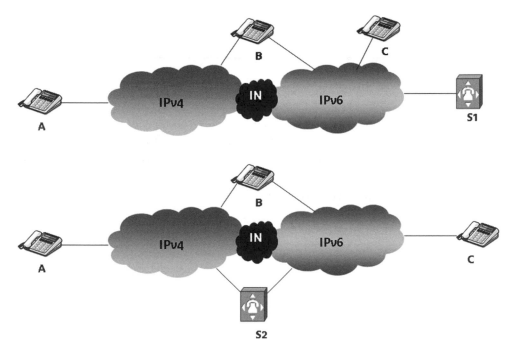

Figure 13.11   IAX in a heterogeneous environment

- **C:** an IAX user agent attached to the IPv6 realm.
- **S1:** an IAX server attached to the IPv6 realm.

In this section, both registration and call setup scenarios are analysed.

13.10.2 Analysis of Registration-Based Operations

This section discusses registration-related messages and therefore registration operations. In particular, it examines whether any hurdles are encountered when IPv6 is activated. The discussions are undertaken from the perspective of a registrar server. Three scenarios are provided:

1. The registrar server is IPv6-only.
2. The registrar server is dual-stack.
3. The registrar server is IPv4-only but the registrant party uses IPv6 to issue its registration request.

The third scenario is not detailed since it is similar to the scenario of an IPv4-only user agent registering in an IPv6-only registrar server.

13.10.2.1 IPv6-Only IAX Registrar

This first scenario supposes an IPv6-only registrar server is deployed. Three use cases are taken into account:

1. An IPv4-only registering user agent (**A**).
2. An IPv6-only registering user agent (**C**).
3. A dual-stack registering user agent (**B**).

In the following, **S1** is used as registrar server, since it is IPv6-only.

IPv4-Only Registrant

Suppose that the IAX registration process is supported by both IPv4-only user agent **A** and IPv6-only registrar server **S1**. In that case, **A** has to send a **REGREQ** request to an IPv4 address (or an FQDN) representing **S1** in the IPv4 world. If an IPv4 address at which to contact the IAX registrar server is provided by the service provider, IAX user agent **A** issues its registration request via an IAX **REGREQ** message to that IPv4 address. If an FQDN to reach the service is provided then a DNS query is sent by **A** to retrieve an IPv4 address. A DNS ALG may be required to translate **AAAA** records (that is, IPv6 records) into IPv4-compatible ones. Doing so, the service provider ensure a global reachability of **S1**. As a result, **A** may send its **REGREQ** to an IPv4 address.

Owing to some routing policies, that **REGREQ** is forwarded to the interconnection node. This element will translate/enforce appropriate operations in order to generate a valid IPv6 packet destined to an IPv6 address of **S1** from the received IPv4 header. In the meantime, an IPv6 address should be assigned at the IN so as to represent unambiguously **A** in the IPv6 realm. If NAT-PT is used, an entry should be instantiated in the NAT table so as to be able to treat subsequent requests and responses. Of course, the service should be engineered so as to translate the destination IPv4 address to a valid IPv6 address representing **S1**. Once a valid IPv6 packet is generated by the IN, it is routed to its destination, represented by an IPv6 address.

Upon receipt of the newly-formed IPv6 packet carrying the IAX **REGREQ**, the IAX registrar server **S1** processes the registration request and stores the perceived IP address and port number (that is, source IP-related information). A **REGACK** message is sent back to the IPv6 address representing **A** in the IPv6 realm. An **APPARENT ADDR IE** is enclosed in **REGACK** to notify the registrant about the perceived IP address from the registrar server's perspective. If the deployed IPv6 transition mechanism is consistent, the **REGACK** message will cross the IN (which should not be understood as a single physical node) and will be delivered to **A**.

At the end of the registration process, **S1** stores an IPv6 address and port number to be used to forward call requests to **A**.

Figure 13.12 illustrates the operations detailed above.

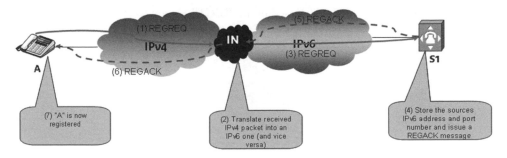

Figure 13.12   Example of a registration call flow (IPv4-only UA in an IPv6-only registrar)

IPv6-Only Registrant

This scenario is similar to the one discussed in Section 13.9. **C** issues its **REGREQ** towards the IPv6 address of **S1**, which stores an IPv6 address at which to reach **C**.

Dual-Stack Registrant

IAX specifications do not handle the dual-stack registrant case. Several alternatives may be adopted:

- *A given dual-stack IAX user agent registers itself within the registrar server using one of its available addresses*: this scenario is the same as an IPv4-only or IPv6-only registration. At the end of the registration process, **S1** stores an IPv6 address at which to reach **C**.
- *The dual-stack IAX user agent issues two registration requests*: the process to send each of these registration requests is the same as for an IPv4-only orIPv6-only registration. The only difference is that the IPv6 registrar server will store two contacts' information to reach this dual-stack IAX registrant. In order to implement this alternative, the IAX protocol specifications need to be enhanced and modified. At the end of the process, **S1** stores two IPv6 addresses at which to reach **C**. The IAX registrar server may maintain only one address of contact.
- *The dual-stack IAX user agent issues one registration request and indicates that it is a dual-stack*: this is another enhancement to the classical behaviour of the IAX protocol. This alternative's philosophy is to indicate to the registrar server the type of this registrant (IPv4-only, IPv6-only, dual-stack). Within this alternative, the registrant will indicate that it is a dual-stack IAX registrant. This indication is denoted within this book as **CONNECTION TYPE IE**. We propose to introduce an information element called **CONNECTION TYPE** which will carry the type of the connectivity supported by an IAX user agent.

This information element is shown in Figure 13.13.
   The value of the **CT** field is equal to:

- **4** in the case of an IPv4-only IAX user agent.
- **6** in the case of an IPv6-only IAX user agent.
- **0** in the case of a dual-stack IAX user agent.

This information element is to be enclosed in the **REGREQ** sent to the IAX registrar server. Upon receipt of this request, the registrar processes it and stores the perceived IP information.

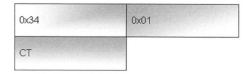

Figure 13.13   CONNECTION TYPE IE

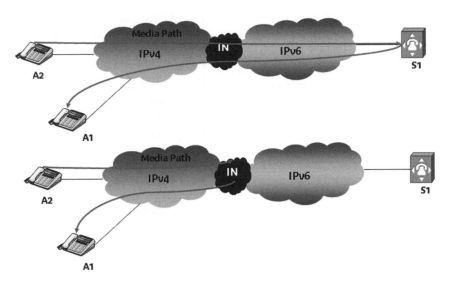

Figure 13.14   Example of a media-optimisation problem

The connectivity type will be used only for call optimisation purposes, especially to optimise the media path.

At the end of the process, **S1** maintains one IPv6 address and an indication of the connectivity type of **CT** (in our case, **CT** is equal to **0**).

Media optimisation is a critical issue when deploying conversational services in a heterogeneous environment. Indeed, if the **CONNECTION TYPE** information element is not supported, the media paths followed by the streams generated by two IPv4-only IAX user agents are not optimal since the perceived address of each user agent is an IPv6 address.

Figure 13.14 shows the media path used to place a call between **A1** and **A2**. In both cases, path-decoupled and path-coupled, the media path is not optimal since it does not follow the shortest IGP (Interior Gateway Protocol) path. If **CONNECTION TYPE IE** was supported, **S1** would be aware that **A1** and **A2** are both IPv4-only user agents and the media transfer process would be slightly modified so as to let both user agents enclose their IPv4 addresses.

Conclusions
Table 13.4 summarises the results of the above discussions.

13.10.2.2 Dual-Stack IAX Registrar

This section focuses on the scenario in which the IAX registrar server is dual-stack.

In the following, **S2** is used as registrar server, since it is dual-stack.

IPv4-Only Registrant
Basic IAX registration procedure applied in this scenario. **A** sends its **REGREQ** request to the IPv4 address of **S2**.

At the end of the process, **S2** stores an IPv4 address to reach **A**.

Table 13.4 IPv6-only registrar considerations

Registrant	IP information stored in the IAX registrar server	Comments
A (IPv4-only)	An IPv6 representation address of **A** in the IPv6 realm	Configuration of the interconnection node should be carefully considered
C (IPv6-only)	Native[1] IPv6 address	None
B (Dual-Stack)	An IPv6 representation address of **B** in the IPv6 realm (Case 1)	This is a non-optimised solution. Discovery means that privileging the IPv6 connectivity should be considered. Impact on the presence of the interconnection node should be taken into account
	Native IPv6 address (Case 2)	None
	An IPv6 representation address of **B** in the IPv6 realm and a native IPv6 one (Case 3)	The advantage of this alternative is to allow call optimisation with other IPv4-only IAX speakers
	An IPv6 representation address of **B** in the IPv6 realm with a connectivity-type indicator (Case 4)	This is a non-optimised solution. Discovery means that privileging the IPv6 connectivity should be considered. Impact on the presence of the interconnection node should be taken into account
	Native IPv6 address with a connectivity-type indicator (Case 5)	Seems to be the recommended solution. But impact on the presence of the interconnection node should be taken into account

[1]Within this chapter, 'native' does not means the IP address configured in a given host but an IP address used within the home IP realm.

IPv6-Only Registrant
Basic IAX registration procedure applied in this scenario. **C** sends its **REGREQ** request to the IPv6 address of **S2**.

At the end of the process, **S2** stores an IPv6 address to reach **C**.

Dual-Stack Registrant
Several scenarios may be envisaged when studying the registration of a dual-stack IAX user agent within a dual-stack registrar server. These scenarios are similar to the ones elaborated in for dual-stack registrants in Section 12.10.2.1.

At the end of the process, **S2** may store:

- One IP address: either an IPv4 address or an IPv6 one.
- Two IP addresses: one IPv4 address and one IPv6 one.
- One IP address (an IPv4 address or an IPv6 one) and an indication about the connectivity type (**CT = 0**).

Table 13.5 Dual-stack registrar considerations

Registrant	IP information stored in IAX registrar server	Comments
A (IPv4-only)	Native IPv4 address	None
C (IPv6-only)	Native IPv6 address	None
B (Dual-Stack)	Native IPv4 address (Case 1)	Call optimisation may be impacted if this alternative is implemented (for calls with IPv6-only nodes)
	Native IPv6 address (Case 2)	Call optimisation may be impacted if this alternative is implemented (for calls with IPv4-only nodes)
	Native IPv4 and IPv6 addresses (Case 3)	The advantage of this alternative is that it allows call optimisation with any IAX user agent
	Native IPv4 address with a connectivity-type indicator (Case 4)	The advantage of this alternative compared to Case 1 is that it allows call optimisation with IPv6-only IAX user agents
	Native IPv6 address with a connectivity type indicator (Case 5)	The advantage of this alternative compared to Case 2 is that it allows call optimisation with IPv4-only IAX user agents

Conclusions
Table 13.5 summarises the results of the above discussion.

13.10.3 Call Setup and Call Optimisation

The registration discussion above shows that in both cases:

1. The registrar server stores an IP address of the same version as its own activated IP protocol.
2. This address can lead to a successful communication with the remote IAX user agent (that is, reception of **REGACK** and **ACK** messages).

Since the placement of an IAX call is similar to the registration procedure, all end-to-end IAX sessions will succeed. In some scenarios, the media path won't be optimised and the intervention of the server will be required.

Below the main conclusions are given regarding call-placement success and call optimisation with an IPv6-only IAX server and a dual-stack one.

13.10.3.1 IPv6-Only IAX Server

Table 13.6 summarises the call setup scenario results and the possibilities of enforcing a call optimisation.

Table 13.6   IPv6-only IAX service call setup and optimisation considerations

Registrant		A	C			B		
				Case 1	Case 2	Case 3	Case 4	Case 5
A (IPv4-only)	Call setup		OK	OK	OK	OK	OK	OK
	Call optimisation		KO	KO	KO	KO	KO	KO
C								
(IPv6-only)	Call setup	OK		OK	OK	OK	OK	OK
	Call optimisation	KO		OK	OK	OK	OK	OK

Figure 13.15 illustrates the path followed by **NEW** messages in the context of a call-placement process. As shown in this figure, all call scenarios are successful and media can be exchanged between heterogeneous IAX peers. The path followed by media streams is not always optimised. Enhancements should be added to IAX so as to allow an optimised media path.

13.10.3.2 Dual-Stack IAX Server

Table 13.7 summarises the call setup scenario results and the possibilities of enforcing call optimisation.

Figure 13.16 provides two call-establishment scenarios to illustrate the behaviour experienced when a dual-stack IAX proxy server is deployed.

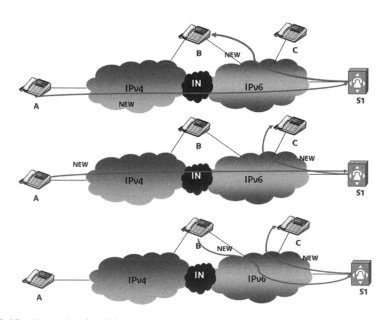

Figure 13.15   Example of call flows observed during call establishment (IPv6-only proxy server)

Table 13.7 Dual-stack IAX service call setup and optimisation considerations

Registrant		A	C	B				
				Case 1	Case 2	Case 3	Case 4	Case 5
A (IPv4-only)	Call setup		OK	OK	OK	OK	OK	OK
	Call optimisation		KO	OK	KO	OK	OK	KO
C (IPv6-only)	Call setup	OK		OK	OK	OK	OK	OK
	Call optimisation	KO		KO	OK	OK	KO	OK

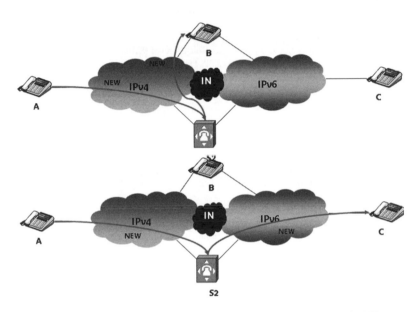

Figure 13.16 Example of call flows observed during call establishment (DS IAX proxy server)

13.11 Conclusion

This chapter discussed the impact of the introduction of IPv6 on IAX services. Several scenarios have been evaluated and discussed. In conclusion, the activation of IAX in an IPv6-enabled environment would not encounter major problems. Note that optimisation methods should be added to IAX specifications so as to optimise the media path followed by streams exchanged between heterogeneous IAX peers.

Service providers should undertake a further effort to identify migration scenarios if IAX is already running in their IPv4 platforms.

References

[BOUCA08] Boucadair, M. and Noisette, N.,'SIP and IPv6', chapter in *SIP Handbook: Services, Technologies, and Security of Session Initiation Protocol*, CRC Press, November 2008.

[CLASSE] Fuller, V., 'Reclassifying 240/4 as usable unicast address space', draft-fuller-240space-02.txt, March 2008.

[HUST] Geoff Huston, 'The IPv4 Report', http://ipv4.potaroo.net.

[RFC4038] Shin, M.-K. et al., 'Application Aspects of IPv6 Transition', RFC 4038, March 2005.

[RFC1883] Deering, S. and Hinden, R., 'Internet Protocol, Version 6 (IPv6) Specification', RFC 1883, December 1995.

[RFC1933] Gilligan, R. and Nordmark, E., 'Transition Mechanisms for IPv6 Hosts and Routers', RFC 1933, April 1996.

[RFC2327] Handley, M. and Jacobson, V., 'SDP: Session Description Protocol', RFC 2327, April 1998.

[RFC3027] Holdrege, M., Srisuresh, M., 'Protocol Complications with the IP Network Address Translator', RFC 3027, January 2001.

[RFC3581] Rosenberg, J. and Schulzrinne, H., 'An Extension to the Session Initiation Protocol (SIP) for Symmetric Response Routing', RFC 3581, August 2003.

[RFC4091] Camarillo, G. and Rosenberg, J., 'The Alternative Network Address Types (ANAT) Semantics for the Session Description Protocol (SDP) Grouping Framework', RFC 4091, June 2005.

[RFC4092] Camarillo, G. and Rosenberg, J., 'Usage of the Session Description Protocol (SDP) Alternative Network Address Types (ANAT) Semantics in the Session Initiation Protocol (SIP)', RFC 4092, June 2005.

[RFC760] Postel, J.(ed.), 'DOD Standard Internet Protocol', Defense Advanced Research Projects Agency, Information Processing Techniques Office, RFC 760, IEN 128, January 1980.

[SIP] Rosenberg, J., Schulzrinne, H., Camarillo, G., Johnston, A., Peterson, J., Sparks, R. et al., 'SIP: Session Initiation Protocol', RFC 3261, June 2002.

Further Reading

A. Conta,'*Extensions to IPv6 Neighbor Discovery for Inverse Discovery*', RFC 3122, June 2001.

B. Carpenter, K. Moore,'*Connection of IPv6 Domains via IPv4 Clouds without Explicit Tunnels*', RFC 3056, February 2001.

B. Haberman, R. Worzella,'*IP Version 6 Management Information Base for the Multicast Listener Discovery Protocol*', RFC 3019, January 2001.

Camarillo, G., El Malki, K., and Gurbani, V., 'IPv6 Transition in the Session Initiation Protocol (SIP)', work in progress, August 2007.

Conta, A. and Deering, S.,'Internet Control Message Protocol (ICMPv6) for the Internet Protocol Version 6 (IPv6)', RFC 2463, December 1998.

Daniele, M., Haberman, B., Routhier, S. and Schoenwaelder, J., 'Textual Conventions for Internet Network Addresses', RFC 2851, June 2000.

Deering, S. and Hinden, R.,'*Internet Protocol, Version 6 (IPv6) Specification*', RFC 2460, December 1998.

Gurbani, V.,et al., 'Session Initiation Protocol (SIP) Torture Test Messages for Internet Protocol Version 6 (IPv6)', RFC 5118, February 2008.

McCann, J., Deering, S. and Mogul, J.,'Path MTU Discovery for IP version 6', RFC 1981, August 1996.

M. Crawford,'*Router Renumbering for IPv6*', RFC 2894, August 2000.

Narten, T., Nordmark, E. and Simpson, W., 'Neighbor Discovery for IP Version 6 (IPv6)', RFC 2461, December 1998.

T. Narten, R. Draves,'*Privacy Extensions for Stateless Address Autoconfiguration in IPv6*', RFC 3041 January 2001.

Thompson, S. and Narten, T., 'IPv6 Stateless Address Autoconfiguration', RFC 2462, December 1998.

14

IAX: Towards a Lightweight SBC?

14.1 Introduction

Session border controllers (SBC, [SBC]) nodes have been proposed, designed and promoted by several vendors (such as Acme, Juniper and so on) so as to meet a set of service provider requirements (both technical and legal). These nodes are not standardised and are proprietary. Several interoperability and service-support issues have been identified by service providers during the validation phase. The introduction of these nodes into operational networks should also be assessed and evaluated from a CAPEX (Capital Expenditure) and OPEX (Operational Expenditure) perspective. Furthermore, the presence of SBC nodes in the service delivery chain introduces additional technical problems and constraints on QoS (Quality of Service) and robustness.

Several of the functions supported by these SBCs are caused by SIP (Session Initiation Protocol, [SIP]) design choices. A lightweight SBC implementation would be envisaged if another signalling protocol was adopted for the delivery of telephony services. From this standpoint, this chapter analyses the functions which would have to be supported by SBC if the IAX protocol was adopted for delivering services.

The chapter is structured as follows:

- Section 14.2 presents the notion of the 'IP telephony administrative domain' and a macroscopic functional view of a telephony service platform.
- Section 14.3 identifies two deployment scenarios for SBC nodes: access and interconnection deployment.
- Section 14.4 provides an overview of the motivations for introducing SBC nodes into SIP architectures. Two categories of motivations are identified and described: technical problems and legal requirements.
- Section 14.5 discusses a set of limitations caused by the presence of SBC nodes.
- Section 14.6 illustrates the functional decomposition of an SBC node. Both media and signalling considerations are described.
- Section 14.7 lists several functions supported by SBC nodes. A brief overview is provided of each.

Inter-Asterisk Exchange (IAX): Deployment Scenarios in SIP-Enabled Networks   Mohamed Boucadair
© 2009 John Wiley & Sons, Ltd

- Section 14.8 checks the applicability of the functions identified in Section 14.7 in an IAX-based service architecture. This section assesses the complexity of SBC nodes when IAX is used instead of SIP.

14.2 IP Telephony Administrative Domain

Service providers administer a set of equipment and service-specific resources such as billing means, authentications procedures, customer profiles databases and so on which interact for the delivery of added-value IP services such as telephony or IP TV. These resources are said to belong to the service provider's domain. In the context of telephony and voice service offerings, this service-specific domain is also denoted 'IP telephony administrative domains (ITAD, [TRIP]). The ITAD delimits the zone covered by the VoIP service provider and includes the voice/media equipment the service provider manages.

Figure 14.1 illustrates several macroscopic functions used to represent a VoIP/ToIP service provider realm in the context of interconnection. This figure shows an ITAD interconnected with several voice/telephony realms such as PSTN (Public Switched Telephone Network), PLMN (Public Land Mobile Network) and other ITADs.

In this macroscopic view, illustrated in Figure 14.1, only a set of functions is represented. For more details regarding a detailed functional architecture, readers are invited to refer to [IMS].

As illustrated in Figure 14.1, an ITAD is composed of several functional elements, as listed below:

- *Proxy server (PS)*: this notion is similar to the one introduced in SIP (Session Initiation Protocol) architectures [SIP]. Within the context of this document, 'proxy server' denotes required functions for the call routing and enhanced services supported by a VoIP service platform. During the signalling phase, a given PS invokes a dedicated element called

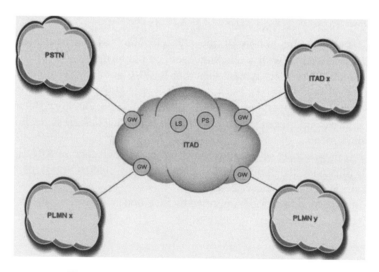

Figure 14.1   IP telephony administrative domain

'location server' (LS) to retrieve the required IP-related information to forward signalling messages in the context of call placement. More information about the location server is provided below.

- *Location server (LS)*: this functional element is responsible for storing the location of customers which are connected to the telephony/voice service. Within this document, the notion of 'location server' is aligned with [TRIP] but also maintains customers' registrations in addition to gateways capabilities. For instance, a location server can be populated by a dedicated protocol such as TGREP (Telephony Gateway Registration Protocol, [TGREP]) or by registration or subscription messages issued from the user agents (UAs) of authorised customers.

- *Gateway (GW)*: this functional element is used to interconnect the ITAD with external voice platforms such as PSTN and other ITADs. Within the context of this chapter, the SBC is considered a GW. A GW intervenes during the media-exchange phase and possibly during the signalling phase when SBCs are present. Within this chapter, this element can also be denoted 'interconnection element' or 'interconnection node'. From a functional standpoint, this element can be separated into two components: SGW (Signalling Gateway) and MGW (Media Gateway). SGWs are responsible for relaying signalling messages between two adjacent realms and MGWs are responsible for the interconnection of two domains and for relaying media message. This functional element is also responsible, when required, for enforcing several functions, such as the ones described in Section 14.7.

14.3 Deployment Scenarios

In current SIP deployments, service providers introduce SBC nodes into their service infrastructure in two service segments, as described in this section.

14.3.1 Access Segment

In order to isolate the service core elements, service providers adopted an engineering rule which is to deploy a back-to-back intermediary node between the customer equipment and the core service elements which host the service logic. Thus, SBC nodes have been proliferated and their deployment accelerated, especially at the access segment. Their deployment was mainly motivated by the need to solve some technical hurdles induced by SIP design choices and to meet a set of legal constraints.

Figure 14.2 provides an example of a telephony domain composed of three SBCs deployed at the access segment. These SBCs are the first contact point visible to end users. Concretely, customers' terminals are provisioned with the required information to send their service data to these elements. An FQDN or an IP address is provided in order to send SIP signalling messages to those elements. Signalling messages are relayed by SBCs to core service nodes. Media streams are exchanged directly between two SBCs without requiring treatment by core service elements.

14.3.2 Interconnection Segment

This section describes the use of SBCs at the interconnection segment (also denoted 'border segment'). First an overview of interconnection scenarios is provided (Section 14.3.2.1) and

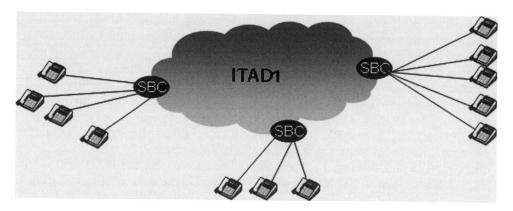

Figure 14.2   Example of SBCs deployed at the access segment

then the utilisation of SBC to interconnect adjacent telephony service realms is described (Sections 14.3.2.2 and 14.3.2.3).

14.3.2.1 Overview

The interconnection between two ITADs is no more than the allocation of access to destination numbers not managed by a single telephony over IP service provider. In other words, the local location server can resolve the locations to which to send a call request, even if this destination is not attached to the local ITAD. To do so, VoIP service providers need to exchange/share the location information, or at least to provide guidelines for the call routing logic of each service provider so that calls to destinations not attached to the local ITAD can be placed successfully. This interconnection of ITADs and exchange of call routing information can follow two approaches:

- *DNS-like approach*: this mode is inspired by the DNS paradigm. In this mode, the collaboration of ITADs in order to offer interprovider VoIP calls does not require any exchange of information between ITADs. Each ITAD configures its DNS server with relevant information for the location of a telephone number. The administrative registration and delegation of number prefixes is not detailed here. In order to place a call, the callee telephony domain must know the called domain server. This is conditioned by the existence of appropriate entries in the DNS system. If no entry exists, the call cannot be initiated.

Figure 14.3 illustrates two ITADs activating a DNS-like interconnection model. In order to process a call request towards **B**, the proxy server of **ITAD1** contacts its LS to ask if **B** is attached to the same ITAD. If not, a query is issued to the global DNS system in order to retrieve the AoC (Address of Contact) of **B**. Note that an address of contact is an IP address and a port number. The AoC retrieved may be that of **B** or that of its inbound proxy server.

Once the location data has been retrieved from the LS, the PS of **ITAD1** forwards its call request towards the AoC. As a result, the call may be established between remote destinations, each of them connected to a distinct ITAD.

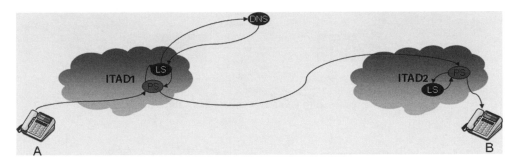

Figure 14.3 DNS-like interconnection approach

- *Flooding approach*: in this mode, two ITADs can peer with each other and exchange adequate information to guide the call routing logic. This can be done statically or by activating a dedicated protocol which aims to propagate/discover and select an ITAD path for call establishment. This mode is similar to the interconnection of autonomous systems at the IP layer. Each ITAD advertises to its adjacent ITADs the prefixes it can reach. This information can be local to an ITAD or based on other information received from other peers.

Figure 14.4 illustrates an example of two adjacent ITADs implementing a flooding-based interconnection approach. **ITAD1** and **ITAD2** are adjacent domains. Both of them deploy service nodes illustrated as PS and LS elements. In this example, **ITAD2** advertises to **ITAD1** its own telephony prefixes. This announcement is stored in LS elements of **ITAD1**, to be used to route call requests towards **ITAD2** prefixes.

14.3.2.2 Use of SBCs

Whatever the adopted interconnection mode, telephony service providers need to control data access to their administrative domains. A set of policies and access rules should be met before access is granted to service resources also in order to prevent DoS (Denial of Service) attacks. For these reasons, and additional ones such as those described in Section 14.7, service providers have deployed and continue to deploy SBC nodes at the interconnection segment. The functions embedded in SBCs located at the interconnection are more or less the same as those of SBCs deployed at the access segment. Functions which are not mandatory for SBCs deployed at the border segment include hosted NAT traversal.

The following subsections provide several configuration schemes for the deployment of an SBC at the interconnection segment.

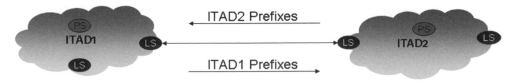

Figure 14.4 Flooding-based interconnection approach

Figure 14.5   Interconnecting two telephony domains through two SBC nodes

14.3.2.3 Interconnection Scenarios

This subsection aims to describe some examples where interconnection is enforced between two adjacent VoIP service realms. The GW node, introduced in Section 14.2, is an interconnection SBC. This node will be responsible for managing both media and signalling at the ingress and egress of a local ITAD.

Several configurations may be considered:

- Each adjacent IP telephony domain deploys its own interconnection SBC node to manage both signalling and media flows, as illustrated in Figure 14.5.
- Both adjacent ITADs share the same interconnection SBC node to interconnect their realms, as illustrated in Figure 14.6.
- Several ITADs share the same element, which may be managed by a third party, as illustrated in Figure 14.7. This dedicated element is denoted within this chapter as a 'telephony exchange point' (TXP).

14.4 Deployment Contexts

The introduction of SBCs into operational networks has mainly been motivated by legal requirements and technical considerations.

14.4.1 Legal Requirements

Telephony services are subject to strict regulatory constraints such as ensuring emergency calls, legal interception and number portability. Additional legal requirements apply in some countries (for example France, where telephony services are part of the 'Universal Services'. Each citizen has the right to ask to be connected to a service regardless of business considerations, such as the expense of laying the last mile).

Figure 14.6   Interconnecting two telephony domains through one SBC node

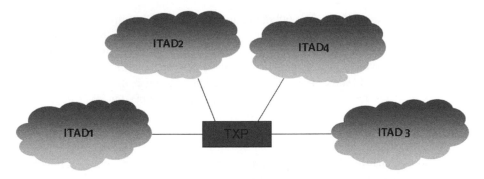

Figure 14.7   Interconnecting several telephony domains through a telephony exchange point

In order to meet some of the aforementioned legal requirements, SBC nodes have been envisaged, and even more deployed, by a plethora of service providers. They have mainly been deployed in order to enforce the following functions:

- *Legal interception*: as described above, SBC nodes are usually positioned at the access segment. Indeed, SBCs are the first 'service' nodes crossed by service flows and are considered a 'concentration' point. SBCs are therefore the place to implement and enforce policies related to legal interception in a transparent manner (that is, without being detected by end users: no service alteration should be noticed, the same traffic profile and service behaviours should be maintained, there should be no QoS degradation and so on).
- *Emergency calls*: since SBCs are the first service contact point reached by traffic issued by the user agent of a given customer, they are the most suitable place to prioritise and enforce emergency services and therefore to ease routing of flows towards a dedicated service platform or PSAP (Public Safety Answering Point).

14.4.2 Technical Considerations

In addition to the legal requirements, SBC nodes have been adopted to meet a plethora of technical hurdles induced by the design of the SIP protocol itself or related to the presence of middleboxes such as firewalls and NATs. For these technical reasons (and additional ones listed in Section 14.7), several vendors have promoted dedicated boxes to be deployed in operational networks.

From an engineering perspective, the introduction of SBC nodes to solve these technical issues is not always motivated, but service providers haven't yet investigated in depth an alternative approach which consists of deploying several lightweight service platforms instead of a single centralised one. These platforms should be located in several geographical PoPs (Point of Presence) and be composed of standardised service nodes instead of SBCs, which are not standardised. Indeed, SBC behaviours are proprietary and each manufacturer implements its own device with close interfaces. Maintenance and management issues should be studied in depth.

The introduction of SBC nodes into operational networks should be taken into account when studying availability and QoS issues (for example, the presence of two nodes in the call path increases the one-way delay). No QoS degradation should be noticed. Note that introducing such a concentration point also introduces a single point of failure. This impacts the robustness and the resilience of the service. The outage of an SBC node might impact a large set of users, who would be out of service (and therefore unable to access the service they had subscribed to through an appropriate SLA (Service Level Agreement)). In order to avoid such failure scenarios, service providers usually deploy backup SBC nodes to secure master SBCs. These backup elements should intervene in the service delivery chain if the master SBC is out of service. This model is denoted as '**2** ***N**' (N being the number of master SBCs deployed in the service architecture). An **N** + **1** backup model has been also investigated, to enhance the robustness of the service platforms. This second model consists of assigning one SBC as a backup. If an SBC in the master pool (an **N** one) fails, this SBC intervenes. Beside this robustness and availability issue, the presence of these intermediary nodes (SBCs) makes it difficult to enforce end-to-end privacy and security mechanisms.

14.5 Service Limitations Caused by SBCs

The presence of SBC nodes in operational networks introduces several (service) limitations to both end users and service providers:

- End users, contrary to the end-to-end SIP spirit, won't be able to benefit from capabilities not supported by those intermediary SBCs. This leads to a hop-by-hop model and not an end-to-end one.

- Service providers require support for new features and/or modification of the configuration of SBC nodes in order to update their service portfolio. Since these nodes are not standardised, the introduction of new services or features might encounter problems if SBC nodes are not updated. This competitive element should be carefully taken into account by service providers, who should privilege standardised and open interfaces with open protocols instead of close platforms and proprietary interfaces.

14.6 Functional Decomposition

An SBC is an intermediary node which is positioned between two realms managed by two administrative entities, as in the interconnection deployment scenario. A given SBC may be dedicated to signalling data and media streams managed by another SBC, or both signalling and media functions may be embedded in the same physical node (see Figure 14.8). For load-balancing reasons, the first scenario may be suitable for deployment.

In the remaining part of this chapter, both media and signalling functions are embedded in the same physical node. Thus, the SBC which handles signalling data will also treat the resulting media streams.

14.7 Taxonomy of SBC Functions in an SIP Environment

SBC boxes have been introduced into SIP architectures in order to meet several service and legal requirements, as elaborated in [SBC]. The objective of this section is not to argue for or against the deployment of these nodes, or to analyse the pertinence of their functions and the concentration of such functions in a single node, but to examine if these functions are still valid if IAX replaces SIP to control VoIP sessions or if they are used at the access or interconnection segments.

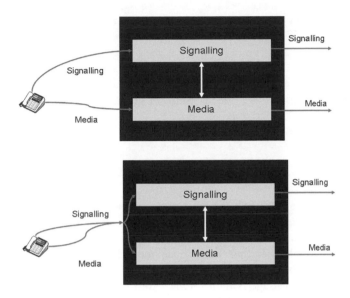

Figure 14.8   SBC functional decompositio

14.7.1 Topology Hiding

Topology hiding consists of limiting the amount of topology information provided to external parties. Service providers do not want to reveal the IP addresses of their service nodes such as proxy servers, registrars and application servers to external parties. This practice is motivated by the need to prevent DoS attacks and not expose engineering polices.

From an SIP perspective, this function acts on SIP headers, which involves stripping Via and **Record-Route** headers, replacing the **Contact** header, and changing **Call-ID**. If IP addresses are used in **From** and **To** headers, a given SBC may replace these with its own IP address or domain name.

In order to illustrate this function, consider Figure 14.9.

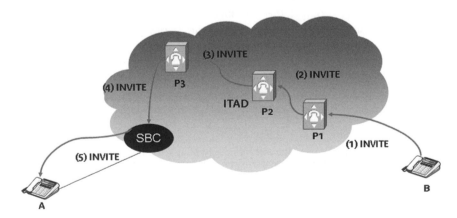

Figure 14.9   Example of the topology-hiding function

Table 14.1   Example of an SIP message before topology hiding is enforced

```
INVITE sip:a@a.domain.example.com SIP/2.0
Via: SIP/2.0/UDP p3.middle.example.com;branch=z9hG4bK48ffgfgfgfg
Via: SIP/2.0/UDP p2.example.com;branch=z9hG4bK18azetddfh
Via: SIP/2.0/UDP p1.example.com;branch=z9hG4bKfjvbghfddfh
Via: SIP/2.0/UDP b.example.com;branch=z9hG4bKdfsdytngny
Contact: sip:b@192.168.2.5
Record-Route: <sip:p3.middle.example.com;lr>
Record-Route: <sip:p2.example.com;lr>
Record-Route: <sip:p1.example.com;lr>
```

In this figure, **B** issues an **INVITE** message to place a call towards **A** (Step 1). This message is relayed by several server proxies (**P1, P2**, and **P3** (Steps 2, 3 and 4)). The **INVITE** message shown in Table 14.1 is then relayed into the SBC.

The output of the topology-hiding function supported by the SBC is shown in Table 14.2.

As illustrated in this example, **Via** and **Record-Route** headers are deleted. The **Contact** header is modified. The message is then forwarded to **A** (Step 5).

This example shows that the internal service topology is hidden from end users. Only SBC nodes are visible to them.

14.7.2 Media Traffic Shaping

Service providers need to control the media traffic exchanged between their attached customers (that is, control the types of RTP flow exchanged in a voice call, avoid covert channels and convey other media types in the context of a call session). SBCs are activated in order to enforce policies based on the types of CODEC used, bandwidth consumption and so on.

14.7.3 Fixing Capability Mismatches

In order to enable session establishment between user agents with incompatible capabilities (CODECs, IP protocol versions and so on), SBCs are inserted in the path so as to fix the capability mismatch. In this context, the introduction of SBCs increases the ratio of successful calls. Of course, SBC nodes should support a large variety of capabilities in order to be able to relay two user agents with incompatible capabilities.

Figure 14.10 illustrates the use of SBC in fixing a capability mismatch. In this example, an SBC node is used for IPv4–IPv6 interworking.

In this example, the SBC relays the message issued by **A** to the **PS** and replaces all IPv4 addresses with IPv6 ones.

Table 14.2   Example of an SIP message after topology hiding is enforced

```
INVITE sip:a@a.domain.example.com SIP/2.0
Via: SIP/2.0/UDP sbc.domain.example.com;branch=z9hG4bKpopopoftdc
Contact: sip:b@192.168.2.5
Record-Route: <sip:sbc.domain.example.com;lr>
```

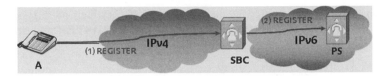

Figure 14.10 Fixing a capability mismatc

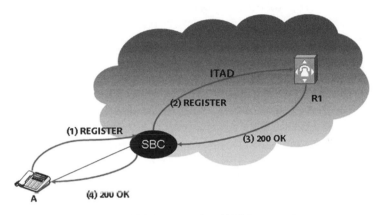

Figure 14.11 Example of NAT traversal

14.7.4 NAT Traversal

A NAT traversal function acts on SIP messages, modifying them so as to allow communication with nodes behind NAT boxes and guess the public IP address to use to contact the NATed clients. Concretely, user agents are configured to issue regular **REGISTER** messages (for example, every **60** seconds). These registration requests are not relayed to core service nodes; a second timer is used for that purpose (for example, a registration timer every **3600** seconds). Adopting this configuration allows NAT entries to be maintained at NAT boxes, so the service platform is not loaded with such traffic.

To illustrate this function, consider a user agent **A** which has issued a **REGISTER** (Step 1, Figure 14.11) message. The message is handled by the intermediary SBC and relayed to the registrar server **R1** (Step 2, Figure 14.11), which issues a **200 OK** message (Step 3, Figure 14.11). The content of this message is shown in Table 14.3.

Table 14.3 Example of a **200 OK** message, as issued by the registrar

```
SIP/2.0 200 OK
From: caller <sip:caller@example.com>;tag=dsfu4
To: called <sip:called@example.com>;tag=sdfs424211
CSeq: 1 REGISTER
Contact: <sips:caller@192.168.1.1>
Expires=3600
```

This response is then transmitted to the originating SBC, which modifies the **EXPIRE** timeout so as to force the originating user to issue regular **REGISTER** requests. Table 14.4 shows the content of the relayed **200 OK** message (Step 4).

Table 14.4 Example of a 200 OK message once treated by the SBC

```
SIP/2.0 200 OK
From: caller <sip:caller@example.com>;tag=dsfu4
To: called <sip:called@example.com>;tag=sdfs424211
CSeq: 1 REGISTER
Contact: <sips:caller@192.168.1.1>
Expires=60
```

14.7.5 Access Control

Service providers which are required to control the type of signalling and media traffic their network carries have an SBC deployed in their operational network. These SBCs are configured so as to control the ingress traffic based on protocol type and additional policies. This function may be supported by classical firewall equipment.

14.7.6 Protocol Repair

A function which repairs protocol messages generated by not-fully-standard clients is supported by SBCs. This function is enabled on SBCs so as to relay to core service nodes only those messages which meet the service traffic profile. Received messages from user agents are checked and modified, if required, so as to be compatible with service requirements (from message format and content perspectives).

14.7.7 Media Encryption

SBCs are used to perform media encryption and/or decryption at the edge of the network. [SBC] argues that this function is required when media encryption is used only on the access network (outer network) side and media is carried unencrypted in the inner network.

14.8 Validity of these Functions in an IAX Architecture

The purpose of this section is to check the applicability of the aforementioned functions in the context of IAX deployment at the access and interconnection segments. Indeed, this section discusses the validity of the aforementioned SBC functions in an IAX environment.

14.8.1 Topology Hiding

Unlike SIP, IAX does not carry information which exposes the internal topology of the service platforms. This function therefore becomes obsolete in an IAX-based service architecture, since no information about either the topology of the service infrastructure or the crossed intermediary nodes is provided.

14.8.2 Media Traffic Shaping

Since IAX is a path-coupled protocol (that is, the media and the signalling path cross the same nodes), this function could be natively supported by IAX servers. Moreover, in the context of IAX and contrary to SIP, the correlation between signalling and media messages is easy to achieve, since the same protocol is used to convey both types of data. Thus there is no need to add additional elements to make this correlation.

14.8.3 Fixing Capability Mismatches

Since IAX is a hop-by-hop protocol (that is, an end-to-end call is decomposed into several call legs and only adjacent IAX peers should have common capabilities), this function becomes obsolete in an IAX context. Furthermore, IAX servers act as relays if remote call participants do not support compatible capabilities.

14.8.4 NAT Traversal

This function is obsolete in an IAX environment since no information about the IP address/port number is enclosed in an IAX message, except for **REGACK** and **TXCNT**. IAX uses the perceived IP information (the source information of a received IP packet) to contact a remote peer. Thus, IAX-based service providers do not need further functions to help NAT traversal. Of course, intermediary NAT binding should be refreshed and maintained. For these reasons, IAX supports native methods such as **PING/PONG**, **POKE/PONG** and **REGREQ/REGACK** to refresh these NAT entries.

14.8.5 Access Control

This function may be obsolete in an IAX context since IAX is a path-coupled protocol and IAX servers can implement this function. Moreover, this function may be supported natively by conventional nodes such as firewalls and other security portals. Only authorized users are able to invoke the service and therefore access service resources.

14.8.6 Protocol Repair

This function may be supported by traditional IAX servers, which modify IAX messages before they are relayed to core service nodes. For instance, if a new IAX client is used by a customer, only the first IAX service node should be able to correctly parse its generated IAX messages. Remote IAX destinations are not aware of this complexity in managing diverse implementations, since this task is achieved by the hop-by-hop mode of IAX (explicitly, by intermediary IAX servers).

14.8.7 Media Encryption

This function becomes obsolete within an IAX context since IAX is a path-coupled protocol. Encryption is always possible between an IAX user agent and the first IAX server in the path. Both IAX peers must use the same encryption scheme.

14.8.8 Lightweight SBC

From a technical standpoint, current SBC implementations would be more lightweight, since the majority of supported functions in an SIP-based environment are not required when IAX is deployed at the access segment (also at the interconnection segment). Some of these functions are supported by native IAX servers (see, for instance, Asterisk implementation) or other conventional service nodes such as firewalls.

14.9 Conclusion

This chapter presented SBC nodes and their supported functions in the context of SIP deployments. These functions are, for the most part, not required if IAX is deployed as the main signalling protocol at the access segment. Thus, investment in these intermediary nodes will be optimised and complexity reduced.

The analysis conducted in this chapter leads to the need for VoIP service providers to check and re-examine the pertinence of mass deployment of SBC nodes in their operational networks. As shown in this chapter, the deployment of SBCs requires an important investment, and in the meantime introduces several additional technical problems, impacting availability, robustness, QoS and load-dimensioning issues. The motivations for introducing SBC into SIP networks are no longer valid in IAX-based service architectures. Traditional service nodes, such as proxy servers and firewalls, should be organised and orchestrated so as to meet engineering requirements.

Service providers should examine and analyse in depth the benefits of introducing IAX in such segments in order to reduce complexity induced by SIP design choices.

References

[IMS] Camarillo, G. and Garcia-Martin, M.A., *The 3G IP Multimedia Subsystem: Merging the Internet and the Cellular Worlds*, John Wiley and Sons, Ltd., 2005.
[SBC] J. Hautakorpi, (ed.), 'Requirements from SIP (Session Initiation Protocol) Session Border Control Deployments', draft-ietf-sipping-sbc-funcs-06, June 2008.
[SIP] Rosenberg, J. et al., 'SIP: Session Initiation Protocol', RFC 3261, June 2002.
[TGREP] Bangalore, M. et al., 'A Telephony Gateway REgistration Protocol (TGREP)', RFC 5140, March 2008.
[TRIP] Rosenberg, J. and Schulzrinne, H.,'A Framework for Telephony Routing over IP', RFC 2871, June 2000.

Further Reading

Acme Packet White Paper, 'Session Border Controllers: Delivering Interactive Communications across IP Network Borders', available at: http://www.acmepacket.com/images/whitepaper_SBC.pdf.
'NAT Traversal for Multimedia over IP Services', available at http://www.newport-networks.com/whitepapers/nat-traversal1.html.
'Solving the Firewall and NAT Traversal Issues for MoIP', available at: http://www.newport-networks.com/cust-docs/33-NAT-Traversal.pdf.

Part Three

Deployment Scenarios in SIP-Based Environments

Part Three is dedicated to elaborating a candidate scenario for the introduction of IAX into an SIP-based environment. It is organised into three chapters:

- *Chapter 15*: advocates for the need to enhance current service architectures and simplify them to avoid complications related to SIP. These complications are induced by SIP design choices, and additional protocols are to be activated to resolve these issues. The activation of those protocols introduces new manageability issues that should be taken into account by service providers when specifying their architectures. Furthermore, this chapter presents the adopted methodology for enhancing the current SIP-based architectures and lists a set of facts to be taken into account. These items should drive the specification effort for an enhancement solution. Some requirements to be considered when proposing new solutions are also described. A brief comparison between IAX and SIP is made. Finally, a set of scenarios in which IAX is activated in operational networks is identified and described.
- *Chapter 16*: provides numerous call flows to illustrate the behaviour of the proposed IAX–SIP interworking function. This chapter shows that the introduction of such a function into operational networks should ease the traversal of middleboxes. It also introduces an extension to SDP to allow end-to-end bandwidth optimisation.
- *Chapter 17*: describes a validation scenario to assess the feasibility of the proposed introduction strategy of IAX into an SIP-enabled environment. This validation scenario does not aim to assess the performance of the proposed solution but only to provide a 'proof of concept' system. Required configuration operations are provided, together with excerpts from configuration files.

15

Scenarios for the Deployment of IAX-Based Conversational Services

15.1 SIP Complications

SIP (Session Initiation Protocol, [SIP]) has been adopted as the main VoIP signalling protocol in various architectures, such as IMS (IP Multimedia Subsystems, [IMS]) and TISPAN (Telecoms & Internet converged Services & Protocols for Advanced Networks, [TISPAN]) architectures. This choice is motivated by the popularity of the protocol and its emergence within the IETF (Internet Engineering Task Force) community. SIP was an answer from the IETF community to the problem of specifying a protocol suitable for controlling multimedia sessions over IP. SIP has naturally been adopted by service providers, owing to its richness and its flexibility. Nevertheless, it suffers from several drawbacks, such as:

1. Complications with crossing NAT (Network Address Translation, [NAT]) boxes.
2. Operational complications with seting up media sessions, due to the dynamic RTP (Real-Time Transport Protocol, [RTP]) port-assignment policy, for instance.
3. Complications due to the path-decoupled nature of SIP. Furthermore, service providers need to insert an intermediate node in both the signalling and the media path, for instance, for access-control purposes.
4. Emergence of SIP-unfriendly boxes which are not standardised and break the SIP end-to-end paradigm.
   The need to deploy a protocol suite close to the famous 'H.323 umbrella':
   - SDP (Session Description Protocol, [SDP])
   - RTP
   - RTCP (Real-Time Transport Control Protocol, [RTP])
   - STUN (Simple Traversal of UDP through NATs, [STUN])
   - TURN (Traversal Using Relay NAT, [TURN])
   - ICE (Interactive Connectivity Establishment, [ICE])
   - and so on.

Inter-Asterisk Exchange (IAX): Deployment Scenarios in SIP-Enabled Networks   Mohamed Boucadair
© 2009 John Wiley & Sons, Ltd

Service providers should take these drawbacks into account in order to investigate how the SIP protocol and companion protocols can be enhanced, or if there are viable alternatives which meet their requirements and do not suffer from these critical 'SIP pains'. The first option (that is, enhancing SIP) is not an easy task, because some SIP complications are caused by its design choices themselves, such as the presence of IP information in the SIP/SDP bodies, which is from an architectural viewpoint a bad practice.

15.2 Structure

This chapter is structured as follows:

- Section 15.3 advocates for the need for an enhancement of current service architectures in order to avoid complications related to SIP. These complications are mainly induced by its design choices. Additional protocols must be activated to solve these issues. But the activation of these protocols introduces new manageability issues that should be carefully considered by service providers.
- Section 15.4 presents the methodology adopted within this chapter to enhance the current SIP-based architectures.
- Section 15.5 lists a set of facts which should drive the specification effort of an enhancement solution. Both financial and technical considerations are listed here.
- Section 15.6 describes some fundamental requirements to be considered when proposing new solutions.
- Section 15.7 briefly compares the ability of IAX and SIP to provide various features.
- Section 15.8 provides a taxonomy of service segments currently deployed by service providers. Three major service segments are described.
- Section 15.9 identifies scenarios to activate IAX in operational networks.

15.3 Beyond the 'SIP-Centric' Era

From the perspective sketched in the previous section, this chapter presents the IAX protocol [IAX] as a possible candidate to solve SIP complications. Indeed, the IAX protocol offers significant features not present in other VoIP signalling protocols. In addition to its simplicity, the main characteristics of the IAX protocol are as follows:

- IAX is transported over UDP (User Datagram Protocol) using a single port number. The default IAX port is **4569**.
- The IAX registration philosophy is the same as the SIP one. An IAX registrant should contact a registrar server with specific messages. Contact information is then retrieved by the registrar server and stored in its system within a time period. The source IP address and source port number are used as the address of contact. No complications are then experienced when contacting a given IAX registrant about a call-establishment request.
- IAX couples signalling and media paths. Decoupling is possible once the connection has been successfully established. This strategy is safe in the sense that the establishment of a successful session is privileged over optimisation or other features.
- Unlike SIP, IAX does not require a new protocol for the exchange of media streams. IAX is also used to send media flows. Numerous media types may be sent by IAX: voice, video, image, text, HTML and so on.

- IAX defines reliable and unreliable messages and does not require TCP or other reliable transport protocols to convey its signalling messages. IAX unreliable messages are mainly media flows which are not acknowledged or retransmitted if lost in the network. IAX reliability is ensured for control messages by several IAX application identifiers maintained by the IAX participants. IAX reliable messages should be acknowledged; if not, they are retransmitted.
- NAT traversal is no longer a nightmare with IAX. No IP addresses are enclosed in the IAX signalling messages issued by an IAX user agent.
- Unlike SIP and RTP, IAX defines a set of messages to monitor the status of the network. These messages can be exchanged during or outside an active call. Pertinent QoS (Quality of Service) indicators are computed. Examples of these are one-way delay, jitter, loss and so on.
- Unlike SIP, IAX offers methods to check whetehr the remote call participant is alive or not.
- Native IP security methods can be deployed jointly with IAX. IAX allows exchange of shared keys. It may be used either with plaintext or in conjunction with encryption mechanisms like AES (Advanced Encryption Standard, [AES]).
- IAX authentication is implemented by an exchange of authentication requests enclosing a security challenge. This authentication challenge should be answered by the remote peer and encrypted according to the adopted encryption method. If encryption negotiation fails, the call should be terminated.
- IAX provides a dedicated scheme to provision IAX devices through IAX messages and a specific procedure.
- IAX allows a procedure to check the availability of a new firmware version for a given device type. The encoding of firmware binary blocks is specific to IAX devices and is out of the scope of the IAX communication protocol itself.

IAX stands as an interesting alternative alongside classical protocols, deployed nowadays by service providers for their conversational service offerings (e.g. H.323 and SIP).

Parts One and Two illustrated that IAX could fulfil a large set of service providers' needs and requirements, and bring still more to their architectures (mainly the native support of traditional services). IAX, as a coupled-path protocol, remains compliant with service provider requirements. Moreover, it provides interesting features such as management of signalling and media transfer, support for native provisioning functions, and firmware maintenance. Chapter 10 sketched how IAX can be enriched with external procedures (such as TRIP [TRIP] and DUNDi [DUNDI]) to enhance the services offered (such as resource discovery, load balancing and so on). Chapter 9 highlighted some interesting features offered by the IAX protocol which should arouse VoIP service providers' interest.

Despite its features richness, IAX is a simple protocol which has the advantage of being IP-version agnostic, leading to avoidance of NAT traversal complications. This issue represents a real asset, as NAT boxes nowadays stand for a tremendous challenge in conversational architectures and services, and require additional patches, especially for home gateway equipment (mainly ALGs) and first service equipment (notably hosted NAT traversal facilities). Moreover, this combination of simplicity and completeness makes it pertinent to avoid resorting to an SIP protocol zoo (SIP, SDP, RTP, RTCP, STUN, ICE and TURN).

15.4 Methodology

To allow a smooth introduction of IAX, an icremental methodology is proposed: first to analytically show the extra value of the IAX protocol over existing ones (this effort is documented in Parts One and Two) and then to propose viable deployment scenarios by which to assess the behaviour of the protocol in operational networks.

IAX can be seen as a complement of service providers' conversational services, for instance at the access segment, or even as a mid-term replacement of the existing protocols.

IAX's native support could help get rid of problems related to NAT: no more heavy ALGs or HNT (Hosted NAT Traversal) mechanisms. This would decrease, if not suppress, the need for expensive SBCs, which moreover aren't required to perform TH (Topology Hiding) operations anymore.

This chapter aims to introduce IAX as a viable solution to operational issues related to the deployment of conversational services. It does not aim to provide detailed specifications of how to enable IAX at access segment, nor to exhaustively identify required functions, but only to propose viable scenarios where IAX capabilities would be of benefit within the operational environment.

15.5 Overall Context

Before we begin to elaborate any IAX-introduction scenarios, several facts should be taken into account.

15.5.1 SIP is Adopted as De Facto Signalling Protocol

SIP has been adopted as de facto signalling protocol by a large set of service providers. This adoption has been motivated by the dynamic created within the IETF around SIP and its associated extensions.

SIP has been promoted as a simple, flexible and extensible protocol. The openness of the protocol has been exploited by protocol designers, who advocate the introduction of SIP to solve any kind of problem. Indeed, proposals have been submitted to use SIP in various contexts, such as establishment of IPSec tunnels, establishment of HIP (Host Identity Protocol) associations, file transfer and so on. To prevent SIP-centric design, the IETF has adopted a 'Request for Comment' [RFC4485] to capture recommendations and guidelines for protocol designers, which will help control SIP extensions. This RFC has been edited by some of the SIP RFC authors.

The danger behind the hegemony of a single protocol is the creation of a meta-protocol that is used in a large set of services and systems. This can lead to complications when evolving these services and systems. The authors of RFC 4485 advocate a modular approach in order to facilitate extensibility and growth; they believe that single protocols could be removed and changed without affecting the entire system.

Additional discussions can be found in [RFC4485].

15.5.2 Heavy Border Equipment

The SIP architecture philosophy, as depicted in [SIP] and reiterated in [RFC4485], is to preserve the independence of signalling and session-related (e.g. RTP stream) packets. This fundamental architectural choice makes SIP a path-decoupled protocol. Nevertheless, this raises several issues when meeting service provider requirements, especially the following:

- *Media traffic shaping*: this function is responsible for controlling the amount of traffic which can be conveyed by a service provider's resources.
- *Fixing capability mismatches*: the purpose of this function is to act as a relay so as to allow successful session establishment between heterogeneous terminals.
- *Maintaining SIP-related NAT bindings*: because of the proliferation of home gateway equipment, NAT is massively used and dynamic bindings must be maintained in order to accept incoming (to the customer site) communication. Dedicated behaviour is adopted by SBCs, mainly by controlling SIP **REGISTER** messages.
- *Access control*: service providers need to control users, protocols and resources in general that benefit from their offered services.
- *Protocol repair*: because SIP specifications are incessantly evolving, not all SIP extensions are supported and 'understood' by all SIP speakers. Such functionality is required to ensure a homogeneous SIP profile and service experience.
- *Media encryption*: in order to ensure secure communications, service providers need to implement security mechanisms which encrypt the media exchanged by their subscribed customers. Because SIP is path-decoupled, an intermediate node must be inserted in the media path. This is achieved by introducing an SBC in both the signalling and the media path. Hop-by-hop security negotiation is then put into effect so as to provide media encryption.

In order to meet these requirements, vendors have promoted new boxes, commonly called session border controllers (SBCs, [SBC]). These products are considered back-to-back user agents (B2BUAs). Both signalling messages and session packets are handled by these new boxes. In addition, SBCs implement functions such as topology hiding to preserve the confidentiality of the service topology and crossed servers. This is required because SIP messages enclose topology information, mainly in **VIA** and **ROUTE** headers. Note that SBC nodes are mostly deployed at access and interconnection points.

Introducing such equipment into operational networks is considered an SIP-unfriendly practice since SIP becomes in practice a path-coupled protocol. The 'flexibility' and 'simplicity' claimed by SIP designers is no longer valid. New functions and protocols are required to allow SIP signals to be routed and delivered to end users.

15.5.3 Routing Confusion

As depicted in [RURI], SIP confuses the notions of 'address', 'name' and 'route'. This is mainly the concern of **REQUEST-URI** and SIP logic to forward based on this parameter. Before forwarding SIP messages, an SIP proxy server must replace the content of the **REQUEST-URI**. This operation induces losing the context of the call, such as obsolescing pertinence of aliases, subaddressing, targeted services and so on.

For further issues related to the **REQUEST-URI** parameter, readers are invited to refer to [RURI].

15.5.4 NAT Traversal Issues

The use of SIP to deliver conversational services has been investigated and solutions have been proposed. Several hurdles caused by the path-decoupled nature of SIP and the interference between the service and network layer have been encountered. In particular, SIP carries information related to the network layer. For these reasons, several problems arise when crossing middleboxes, mainly NAT and firewalls. Plenty of solutions have been introduced, such as:

- UPnP (Universal Plug and Play)
- STUN (Simple Traversal of UDP through NATs)
- Connection-Oriented Media
- Symmetric RTP
- TURN (Traversal Using Relay NATs)
- Media Relay (combination of symmetric RTP and TURN server)
- ICE (Interactive Connectivity Establishment)
- And so on.

When deploying these solutions, manageability issues should be taken into account.

15.5.5 The Need to Take into Account the IP Exhaustion Problem and Migration to IPv6: Easing IPv4–IPv6 Interworking

Based on simulation studies conducted by G. Huston, the date when IPv4 addresses will become unavailable is December 18th, 2009. This date may be seen as a catalyst to seriously kick off IPv6 deployment in operational networks. For this purpose, and in order to drive the process, ARIN (American Registry for Internet Numbers) has adopted a resolution which mandates the promotion of IPv6 usage. This rule must be followed in order to request new IP address blocks.

As a short-term solution, some service providers have investigated and even deployed an alternative which optimises the required public IP addresses for the delivery of added-value IP service offerings. This solution proposes to introduce an additional level of NAT, denoted 'Provider NAT' (also called 'Double NAT'). This second level of NAT is hosted at the service provider perimeter. This solution assumes that no public IP addresses are assigned to end users. Only private IP ones are assigned to end users (more precisely, to their equipment). When the traffic issued by end-user terminals needs to exit the service provider private network, an NAT operation is required at the Provider NAT box.

When delivering SIP-based services with Provider NAT nodes deployed, these constraints should be considered:

- The service provider should be aware of the underlying IP infrastructure so as to implement appropriate ALGs (Application Level Gateways). At least two modifications of SIP messages should be applied: the first at the Home NAT and the second at the Provider NAT. If no such ALG is enabled, no communication may be established. This constraint is 'heavy', since it assumes a vertical integration (that is, no functional separation between the service provider and the IP network provider) and that the same administrative entity administers both service and network infrastructure.
- NAT mapping entries at the Provider NAT box should be maintained so as to be able to deliver incoming message to customer devices located behind the Provider NAT.

- Media flows may encounter some problems with their delivery, since RTP ports may not be opened.

As a consequence, if a Provider NAT solution is deployed, SIP-based services may be impacted and their quality of experience altered.

Beside this short-term solution, in the long term, as far as conversational services are concerned, especially SIP-based services, service providers must undertake activities to solve SIP complications with regards to IPv6 and interworking with IPv4 realms. From this standpoint, IAX may be seen as a viable alternative able to set successful multimedia communication between heterogeneous nodes. For more information, the reader is invited to refer to Chapter 13.

15.5.6 Fixed–Mobile Convergence

Most service providers have identified fixed–mobile convergence as a major target. The motivations are both technical and financial. From this standpoint, both mobile and fixed terminals should be taken into account when designing IP-based services. More precisely, mobile devices and handsets should be able to access SIP-based services. Therefore, SIP stacks requiring less memory must be implemented in these devices.

Studies to compare the memory required for SIP and alternative protocols such as IAX should be undertaken.

15.5.7 Investment

Important investments have been made by major service providers to replace their PSTN offerings. These investments should be taken into account when investigating new technical tracks to enhance service offerings and ease their deployability. This requirement is critical and is outside the scope of this book.

15.5.8 Implementation Availability

Besides the emergence of SIP in standardisation fora, IAX has been adopted by the open-source community as the main protocol to allow VoIP trunk exchange between servers. Thus, several service offerings have been introduced using IAX. A list of service providers that have deployed IAX can be found at www.voip-list.com/protocols/iax_1.html or www.voipcharges. com/providers/country_all/iax2/all. At least 151 service providers deploy IAX.

Examples of IAX-based service offerings are Jajah (available at www.jajah.com) and VoipBuster (available at www.voipbuster.com/fr/index.html).

IAX has been promoted within the Asterisk community (www.asterisk.org), maintaining an open PBX (Private Branch Exchange). Moreover, products, such as gateways, which support IAX are available (e.g. the VoIP GW-200 IAX2 (www.neodiscount.com/voix-sur-ip/ telephonie-ip/Passerelles_IAX.htm)). IAX hardphones are also available, for example the ST-302 (www.asteriskguru.com/tutorials/st_302_ip_phone_hardphone.html) and the Atcom AT-320PD IAX2 (accesip.fr/product_info.php?products_id=5).

15.5.9 Conclusion

SIP suffers from some critical problems, which require the introduction of additional functions and equipment. These technical issues can be avoided by the adoption of one of two strategies: modifying SIP protocol design or investigating an alternative protocol.

Within this book, the focus is on the second track.

15.6 Architectural Requirements

Within the context of protocol and/or architecture migration, the following requirements should be met by any proposal to deliver conversational services:

- *Backward compatibility*: new adopted architectures and associated protocols must be backward compatible. This requirement aims at evaluating the risk and impact of deploying a new solution on the infrastructure already in place. New solutions should provide adequate guarantees of backward compatibility, not only to allow a smooth migration, but also to prevent existing infrastructures from becoming unstable. New solutions should also allow smooth migration, to enhance the service and allow advanced services over the solution already in place.
- *Migration transparency and service continuity*: services perceived by end users should not be altered. The migration of service platforms and introduction of new protocols should be transparent for end users.
- *Protocol modularity*: as highlighted in [RFC4485], protocol modularity should be preserved. Thus, specific protocols for specific technical issues should be preferred to a single 'do all' protocol.
- *A path-coupled model at service provider domains' boundaries*: service providers should be able to control both the signalling and the media path. This requirement is motivated by the need to enforce traffic shaping, access control, encryption and so on. It should be valid at least at the access and interconnection segments.
- *Ease of service maintenance and provisioning*: the manageability of new protocols should be assessed and validated before they are introduced into operational networks. This feature should be easy to enforce and put into effect.
- *Solution to middlebox complications*: any new alternative to SIP must be able to cross NATs and firewalls. This requirement is fundamental since this is one of big hurdles of SIP implementations that require the activation of companion protocols such as ICE, STUN or TURN.

15.7 Brief Comparison

This section provides a concise comparison between SIP and IAX. The purpose is to identify where IAX simplifies the establishment of multimedia sessions over SIP. There is also a focus on areas where SIP should not be replaced.

15.7.1 Signalling Message Length

In order to illustrate the difference between the length of an IAX control message and an SIP message, consider the following scenario, where **Med** calls **Pierrick Morand** using SIP and

Table 15.1 INVITE message

```
INVITE sip: 835262@192.168.1.2; SIP/2.0
Via: SIP/2.0/UDP 192.168.1.1:5060;branch=z9hG4bKfds541254sdfdf
From: "Med" <sip:832654@192.168.1.1>;tag=qsdf5412
To: <sip:835262@192.168.1.2>
Call-ID: dfsdf45sdf21sqdf5412qsdf54@orange
CSeq: 1 INVITE
Max-Forwards: 70
Contact: <sip:832654@192.168.1.1:5060>
User-Agent: linphone
Accept-Language: en
Accept: application/sdp
Allow: INVITE, ACK, CANCEL, BYE, REFER, OPTIONS, NOTIFY, SUBSCRIBE, PRACK,
MESSAGE, INFO
Allow-Events: talk, hold, refer
Supported: timer, 100rel, replaces
Session-Expires: 3600
Proxy-Authorization: Digest
Content-Type: application/sdp
Content-Length: 283
v=0
o=pmo 1158699929 1158699929 IN IP4 192.168.1.1
s=dfdvdsvdv
c=IN IP4 192.168.1.1
t=0 0
m=audio 12345 RTP/AVP 0 8 3 18 101
a=rtpmap:0 pcmu/8000
a=rtpmap:8 pcma/8000
a=rtpmap:3 gsm/8000
a=rtpmap:18 g729/8000
a=rtpmap:101 telephone-event/8000
a=fmtp:101 0-15
a=sendrecv
```

IAX. The length of an SIP message depends on the conveyed capabilities; it may be more than **1000 bytes**. IAX ones are much smaller (close to **60 bytes**), as illustrated below:

- Table 15.1 gives an example of an SIP **INVITE** message (not all optional lines are included).
- Table 15.2 gives an example of an IAX **NEW** message (this trace is not binary).

15.7.2 Media Stream Length

SIP is used to initiate multimedia sessions. RTP is then used to exchange media traffic. RTP flows enclose at least **12 bytes** of overhead per packet if no CSRC (Contributing Source) is included. As for IAX, only **4 bytes** of overhead are required to carry media traffic. Figure 15.1 shows the difference between an IAX mini frame and an RTP one.

Table 15.2 IAX **NEW** message

```
Rx-Frame Retry [No] OSeqno: 000 ISeqno: 000 Type: IAX Subclass: NEW
Timestamp: 00000ms SCall: 211221 DCall: 00000 [192.168.1.1:4569]
VERSION: 2
CALLED NUMBER: 835262
CALLING NAME: Pierrick Morand
USERNAME: EVI_RURD
FORMAT: 4
CAPABILITY: 4
```

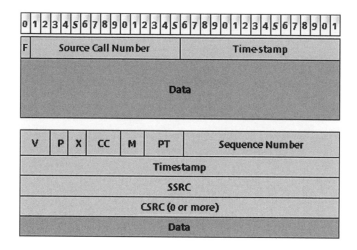

Figure 15.1 IAX vs RTP

IAX can also be activated with a trunk mode. The bandwidth optimisation ratio compared to RTP is important. To illustrate this mode, consider an example with the G.729 CODEC (8 Kbps bit rate and payload size 20 ms (50 packets per second)).

- Figure 15.2 provides the formula to compute the required bandwidth when RTP is used to exchange media streams. For example:
  - for **n = 50, BW = 1.48 Mb/s.**
  - for **n = 100, BW = 2.96 Mb/s.**

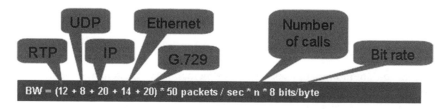

Figure 15.2 RTP bandwidth formula

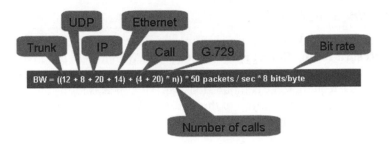

Figure 15.3   IAX bandwidth formula

- Figure 15.3 provides the formula to compute the required bandwidth when IAX is used in a trunk mode. For example:
  - for **n = 50, BW = 0.50 Mb/s**
  - for **n = 100, BW = 0.98 Mb/s.**

15.7.3 Security

Both IAX and SIP support security mechanisms to achieve encryption and authentication. For media encryption, SIP relies on TLS (Transport Layer Security) and SRTP (Secure RTP), and IAX on AES-128. For authentication, both SIP and IAX use MD5 [MD5]. Due to its end-to-end nature, SIP security negotiation may fail because of the intermediate node. Additional procedures to ease the traversal and processing of ciphered SIP messages have been investigated and proposed within IETF.

15.7.4 NAT Traversal

In order to cross NAT boxes, SIP relies on several protocols, such as STUN, TURN and ICE. Because of its design choices, IAX avoids problems related to NAT. These design choices are: use of UDP, use of a single UDP port and not carrying any IP addresses in a message.

15.7.5 Peer-to-Peer

Because of its difficulties with crossing NAT boxes, SIP should encounter difficulties when used in a P2P (Peer-to-Peer) architecture. At least two relays will be required: one for the signalling message and one for media streams.

15.7.6 Firewall Traversal

Because a dynamic port number assignment policy is adopted in SIP-based communication, RTP flows may encounter problems passing through firewalls. Additional techniques must be activated in order to allow these communications to be accepted by intermediate firewalls. IAX does not encounter this issue, since it uses the same port number for both signalling and media streams. This port is well known and an appropriate rule may be configured in firewall equipment in the same way as for HTTP traffic.

15.7.7 Routing Considerations

Several headers must be taken into account in order to answer an SIP message: **MADDR, VIA (received), Record-Route, Route, rport** and **Contact**. Within IAX, there is no such confusion; the reply is sent the source of the message. Moreover, because SIP uses the **REQUEST-URI** to forward the request, some complications can arise, such as losing the targeted service, aliases and so on. A complete analysis of this hurdle is documented in [RURI]. IAX does not suffer from such complications.

15.7.8 IPv6–IPv4 Interworking

The SIP protocol and architectures must be modified in order to set successful calls between nodes located in IPv4 and IPv6 realms. This complication is not encountered in IAX. If IP mechanisms are available to interconnect IPv4 and IPv6 realms, an IAX session can be established without major pains.

15.7.9 Keep-Alive Feature

SIP does not support any feature to test the activity of a remote peer. Therefore, an SIP speaker cannot know whether or not a remote peer is still in the session. IAX introduces a scheme based on the exchange of specific messages, called **PING/PONG**, which aim to check whether remote peers are still alive. A dedicated timer is used to tune the exchange of such messages.

15.7.10 Forking

IAX does not support forking. This feature is important in situations where several 'next hops' can be contacted in order to place a call. SIP supports forking capabilities.

15.7.11 Routing

SIP uses several headers, such as **Route, Record-Route** and **Via**, to ensure symmetry in signalling paths. IAX does not carry such information in its signalling messages. Additional tables are therefore required to implement stateful call processing. SIP also requires these state tables.

15.8 Taxonomy

A service platform can be divided into three service segments, as represented in Figure 15.4.

15.8.1 Access Segment

This service segment includes functions which are required for connecting customer equipment to the service-offering platform. It may include DHCP servers, DHCP client relays and so on. It is usually represented as 'POP' (Point of Presence). Equipment dedicated for the use of geographically-distributed customers is dispersed between PoPs. These POPs are connected to the core segment via dedicated interfaces. SIP complications are experienced to a large extent in these segments.

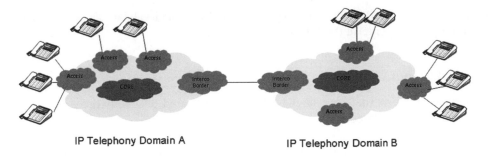

IP Telephony Domain A   IP Telephony Domain B

Figure 15.4   Service segments

15.8.2 Core Segment

This service segment is the 'hearth' of the service platform. It is where the service logic and required functions such as routing, billing and so on within IMS architecture are hosted. It is also responsible for interconnecting internal or external ASs (Application Servers).

15.8.3 Border Segment

This segment groups functions required to interconnect with external realms. These external realms may be VoIP, PSTN (Public Switched Telephone Network), PLMN (Public Land Mobile Network) or any other service domain. This service segment is important because financial data depends on the records it collects. In order to preserve confidentiality and hide the service topology, embedded functions should support THIG (Hiding Inter-working Gateway) and some SIP headers must be modified/hidden/dropped. Alongside the access segment, this segment suffers from some complications related to use of SIP as a signalling protocol.

15.9 Introducing IAX into Operational Networks

15.9.1 Rationale

As stated above, complications related to SIP-based service platforms are experienced in both access and border segments, especially with crossing middleboxes and enforcing a path-coupled scheme. Additional complications have been elaborated in previous chapters. Meanwhile, IAX is an interesting protocol which is compact, simple and easily able to cross middleboxes, maintain a session's aliveness and so on; IAX can therefore be activated in the aforementioned segments to avoid most of the complications encountered by SIP. The position of this chapter is not to argue in favour of replacing SIP with IAX, because this is not a viable solution; instead, it argues in favour of introducing IAX only in situations where it is better than SIP, and maintain SIP for other uses where it is more appropriate.

The position is not the same for service providers that have already deployed IP-based telephony services. For these service providers, IAX should be introduced in a transparent and incremental manner. IAX should be activated in accordance with SIP. An example of an SDP extension is described in Chapter 16 to illustrate how SIP can be used to manage IAX-based media sessions. For service providers that did not deploy an IP-based telephony service, IAX can be envisaged as a concurrent solution to SIP. For these service providers an in-depth analysis should be undertaken to compare SIP and IAX and validate an optimal solution regarding plenty of criteria (both technical and financial).

15.9.2 Alternatives

For service providers that have already deployed SIP-based telephony services, two alternatives may be considered:

- *Replace SIP with IAX only at the access segment*: this scenario consists of activating IAX to access the service from customer sites. SIP is maintained in the core and border segments. A dedicated gateway to interconnect IAX with the SIP-based world must be introduced. This scenario is elaborated in Chapter 16.
- *Replace SIP with IAX in both the access and the border segments*: in this scenario, both the access and the border segments are implemented by the activation of IAX. By its nature, IAX meets most service provider requirements, such as topology hiding, path-coupled architecture, control of both media and signalling messages, media encryption support and so on.

An additional scenario may be envisaged, where the experienced service is enhanced and the bandwidth use optimised by replacing SIP even at the core segment. Within this book we focus on the first scenario; both the second and third scenarios are outside the scope of this book.

15.10 Conclusion

This chapter presented the IAX protocol as a possible candidate to solve SIP complications. The rationale is not to replace SIP with IAX but to adopt a 'win-win' strategy in which both IAX and SIP are used to meet service provider requirements. In particular, this chapter recommended using IAX in service segments in which SIP encounters technical complications. Introducing IAX in these segments eases the manageability of the deployed architecture in terms of required protocol. No 'SIP protocols umbrella' is then required, since IAX solves the aforementioned technical hurdles. IAX can also be used in an SIP-based environment to convey media streams and therefore optimise the use of bandwidth. In order to implement this mode, SDP and SIP should be extended to carry IAX-related information in the SDP offers.

Finally, this chapter advocated for the activation of IAX at the access segment. A detailed description of this is provided in Chapter 16.

References

[AES] US Department of Commerce/NIST, 'FIPS-197, Announcing the Advanced Encryption Standard', November 2001.
[DUNDI] Spencer, M., 'Distributed Universal Number Discovery (DUNDi)', draft-mspencer-dundi-01, October 2004.
[IAX] Spencer, M., Shumard, K., Capouch, B. and Guy, E., 'IAX2: Inter-Asterisk eXchange Version 2', draft-guy-iax, work in progress.
[ICE] Rosenberg, J., 'Interactive Connectivity Establishment (ICE): A Methodology for Network Address Translator (NAT) Traversal for Offer/Answer Protocols', draft-ietf-mmusic-ice-12, October 2006.
[IMS] Camarillo, G. and Garcia-Martin, M.A., 'The 3G IP Multimedia Subsystem: Merging the Internet and the Cellular Worlds', John Wiley and Sons, Ltd., 2005.

[IPSEC] Kent, S. and Atkinson, R.,'Security Architecture for the Internet Protocol', RFC 2401, November 1998.

[MD5] Rivest, R.,'The MD5 Message-Digest Algorithm', RFC 1321, April 1992.

[NAT] Holdrege, M. and Srisuresh, M., 'Protocol Complications with the IP Network Address Translator', RFC 3027, January 2001.

[RFC4485] Rosenberg, M. and Schulzrinne, H., 'Guidelines for Authors of Extensions to the Session Initiation Protocol (SIP)', RFC4485, May 2006.

[RTP] Schulzrinne, H., Casner, S., Frederick, R. and Jacobson, V., 'RTP: A Transport Protocol for Real-Time Applications', RFC 1889 (proposed standard), January 1996.

[RURI] Rosenberg, J.,'Applying Loose Routing to Session Initiation Protocol (SIP) User Agents (UA)', draft-rosenberg-sip-ua-loose-route-01, June 2006.

[SBC] Hautakorpi, J. et al., 'Requirements from SIP (Session Initiation Protocol) Session Border Control Deployments', draft-camarillo-sipping-sbc-funcs.

[SDP] Handley, M., Jacobson, V. and Perkins, C.,'SDP: Session Description Protocol', RFC 4566, July 2006.

[SIP] Rosenberg, J., Schulzrinne, H., Camarillo, G., Johnston, A., Peterson, J., Sparks, R. et al., 'SIP: Session Initiation Protocol', RFC 3261, June 2002.

[STUN] Rosenberg, J., Weinberger, J., Huitema, C., and Mahy, R.,'STUN: Simple Traversal of User Datagram Protocol (UDP) Through Network Address Translators (NATs)', RFC 3489, March 2003.

[TISPAN] TISPAN, 'Telecommunications and Internet converged Services and Protocols for Advanced Networking, NGN Release 1', TR180001, 2006.

[TRIP] Rosenberg, J. et al., 'Telephony Routing over IP (TRIP)', RFC 3219, January 2002.

[TURN] Rosenberg, J. et al., 'Traversal Using Relay NAT (TURN)', work in progress.

16

IAX in the Access Segment of SIP-Based Service Architectures

16.1 Introduction

The main purpose of this chapter is to describe a suitable scenario for the integration of IAX into operational service platforms. The motivation behind this is as follows:

- IAX does not suffer from the drawbacks of existing VoIP signalling protocols such as SIP.
- IAX is compatible with the path-coupled philosophy adopted by service providers, especially at the access segment (rational for introducing SBC nodes).
- SIP is rich in terms of routing and forking capabilities and also its native support of path-decoupled schemes suitable for intraprovider concerns.

The scenario proposed in this then chapter is to activate IAX at the access segment and maintain SIP in the core, as shown in Figure 16.1.

For this purpose, IAX–SIP and IAX–RTP gateways should be deployed at the access segment so as to translate IAX messages into either SIP or RTP (Real - Time Transport Protocol, [RTP]) messages, as illustrated in Figure 16.2.

The purpose of this chapter is not to specify such gateways in depth but only to describe their behaviours at a conceptual level. Therefore, the following sections provide various call flow examples illustrating their roles.

16.2 A 'High-Level' Description of the Interworking Function

Figure 16.3 sketches a subset of the functional architecture of an IAX–SIP interworking function (IWF). This function is responsible for:

- Translating received IAX control message into SIP ones.
- Translating received media IAX messages into RTP ones.
- Translating received SIP messages into IAX control ones.
- Translating received RTP messages into IAX media ones.

Inter-Asterisk Exchange (IAX): Deployment Scenarios in SIP-Enabled Networks   Mohamed Boucadair
© 2009 John Wiley & Sons, Ltd

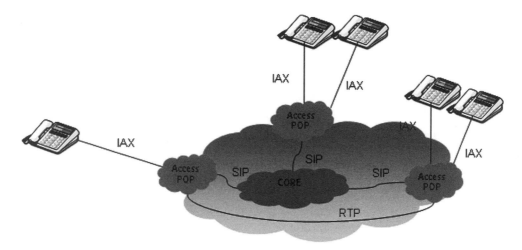

Figure 16.1 IAX at the access segment of conversational services

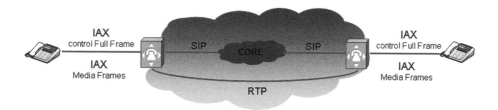

Figure 16.2 IAX gateways

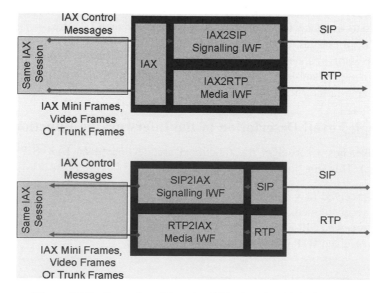

Figure 16.3 Examples of functional blocks of an IAX–SIP IWF

This IAX–SIP interworking function is deployed in the first access node in the access part of a given telephony administrative domain. It supports the following clients:

- **IAX Client** at the inner interface (connecting end devices to service platform).
- **SIP Client** at the outer interface (connected to service core nodes).
- **RTP Client** at the outer interface (connected to service core nodes).

When an IAX message is received from an IAX user agent, that message is handled by the **IAX Client**, which classifies the message according to its type:

- If the received message is an IAX control message then it is passed to **IAX2SIP Signalling IWF**.
- If the received message is an IAX media message (a mini frame or a video full frame, for example), the message is sent to **IAX2RTP Media IWF**.

As an output:

- **IAX2SIP Signalling IWF** generates an SIP message.
- **IAX2RTP Media IWF** generates an RTP message.

When an incoming message is received from a core service node, that message is handled by either the **SIP Client** or the **RTP Client**.
    As an output:

- **SIP2IAX Signalling IWF** generates an IAX control message.
- **RTP2IAX Media IWF** generates an IAX media message.

The IAX–SIP interworking function is stateful and maintains states related to ongoing sessions.
    The next section provides several call flows to illustrate the behaviour of the IAX–SIP interworking function.

16.3 Examples of Call Flows

16.3.1 Reference Architecture

This section is dedicated to illustrating interworking scenarios when IAX is deployed at the access segment. In this context, SIP is activated to interconnect core service nodes. The below examples are elaborated taking into account the architecture illustrated in Figure 16.4. In this figure, the following elements are shown:

- **A** and **B** are two terminals used by two remote users. These users are subscribed to the telephony service offering.
- **GW1** and **GW2** are two access nodes which host an IAX–SIP interworking function.
- **PS** is an SIP-based proxy server.
- **R** is an SIP-based registrar server.

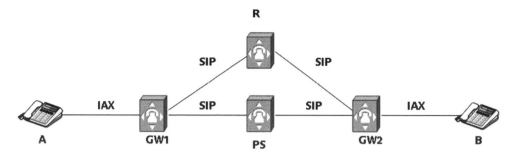

Figure 16.4 SIP–IAX reference architecture

In the architecture represented in Figure 16.4:

- IAX is used to carry both signalling and media streams between **A** (respectively **B**) and **GW1** (respectively **GW2**).
- SIP is used as a main communication protocol between **GW1** and **PS** (respectively **GW1** and **R, GW2** and **PS,** and **GW2** and **R**).

Figure 16.4 may be simplified to use the same physical node to host the registrar server and the proxy server. This option is not taken in this chapter. Furthermore, interfaces required to store the location information and customer profile database are not represented in this figure for simplicity reasons. A given service provider may deploy an IMS-based architecture in its core segment.

The below call flows should not be understood as recommended specifications but only as examples to implement an IAX–SIP interworking function.

The following scenarios are detailed in this section:

- Use of IAX for firmware update and provisioning is described in Section 16.3.2.
- Registration without authentication is described in Section 16.3.3.
- Registration with authentication is described in Section 16.3.4.
- Some call setup scenarios are given in Section 16.3.5.
- A call tear-down scenario is given in Section 16.3.6.
- Registration release without authentication is described in Section 16.3.8.
- Registration release with authentication is described in Section 16.3.9.
- Section 16.4 introduces a new procedure to optimise media exchange between access nodes.

16.3.2 Provisioning and Firmware Update of End Devices

When IAX is used at the access segment, native IAX capabilities should be used and exploited. As an example, provisioning and firmware update of end-user devices should be activated and implemented. Since SIP does not support such means, no interworking issue is raised. Consequently, the procedure described in Chapter 8 can be activated.

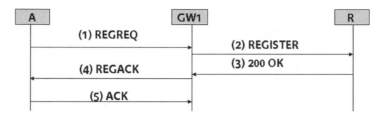

Figure 16.5   Example of a registration without authentication call flow

16.3.3 Registration without Authentication

Figure 16.5 sketches a simple example of subscription to a telephony service when an IAX–SIP interconnection node is deployed in the service platform. This example assumes that no authentication procedure is activated by the service provider. It is not recommended and is provided only for illustration purposes.

Furthermore, this section assumes that the registration is ensured by an external node rather that the interconnection node itself.

As shown in Figure 16.5, the following messages are exchanged between **A** and the service nodes so as to let **A** register with the service:

- An IAX **REGREQ** is sent by **A** to the first service contact point. In this example, the first service contact point of **A** is **GW1**. An IP address or an FQDN (Fully Qualified Domain Name) was provided to **A** prior to the registration process. Several methods may be used to provide such configuration information.
- The **REGREQ** message is then received by **GW1**. The IAX–SIP interworking function is invoked. As a result, a **REGISTER** message is built. This message reuses identity information related to A. **GW1**'s headers may be **CONTACT** and **EXPIRES**.
- Upon receipt of the **REGISTER** message, **R** retrieves both the AoR and the AoC of **A**. A **200 OK** is then sent back to **GW1**.
- Once received by GW1, the SIP–IAX interworking function is invoked and a **REAGACK** is built and sent to **A**.
- As a response, an **ACK** message is sent to **GW1**. At this stage, **A** is registered with the service and may place and receive calls to/from remote user agents.

16.3.4 Registration with Authentication

Unlike the previous registration example, Figure 16.6 shows the message exchanges that occur when authentication procedures are activated by a given service provider. In this scenario, service providers ensure that a remote user agent is allowed to issue a given control message. Authentication procedure may be implemented at the interconnection node (e.g. **GW1**) or at the core node (e.g. **R** or another external node). In this section, the second scheme is assumed. In that case, access nodes relay received messages to the core service node to invoke authentication procedures.

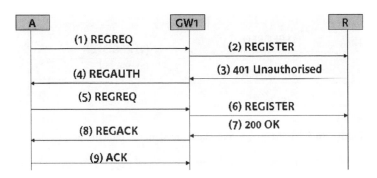

Figure 16.6   Example of a registration with authentication call flow

As represented in Figure 16.6, the following steps are taken to subscribe to a service:

- As in Figure 16.5, a **REGREQ** and an associated **REGISTER** message are generated and sent to their destinations.
- Once the **REGISTER** is received by **R**, authentication methods are invoked. As a result, **R** sends a **401 Unauthorized** message to **GW1**.
- The SIP–IAX interworking function is invoked. A **REGAUTH** is generated and sent to **A**. This **REGAUTH** message encloses a security challenge.
- Upon receipt of the **REGAUTH**, **A** computes a security hash using the received security challenge. This hash is included in a second **REGREQ**, which is routed to **GW1**.
- Once the **REGREQ** is received by **GW1**, the IAX–SIP interworking function is invoked. As a result, a **REGISTER** message is built. This message includes a security hash. The **REGISTER** is then sent to R.
- R proceeds to verification operations to check if the security hash is valid according to the authentication methods put in place. To acknowledge this treatment, a **200 OK** message is sent back to **GW1**.
- The interworking function generates a **REGACK** message and forwards it to **A**.
- As a final step, **A** sends an **ACK** message. When this message is received by **GW1**. A is registered with the service and there is no need to reissue any REGACK messages.

16.3.5 Call Setup

16.3.5.1 Basic Procedure without Authentication

Figure 16.7 illustrates the exchange of messages that occurs during a call setup between **A** and **B**. In this figure, m ini frames exchanged for ring tones are not represented.

It is assumed that a direct communication can be established between **GW1** and **GW2** using RTP. Section 16.4 introduces a novel solution which optimises the bandwidth required to exchange media traffic between **GW1** and **GW2**.

Concretely, in order to place a call between **A** and **B**, these steps are followed:

- A **NEW** message is issued by **A** to initiate a call to **B**. This message is sent to the service contact point. In this example, the first service contact point of **A** is **GW1**. An IP address or an

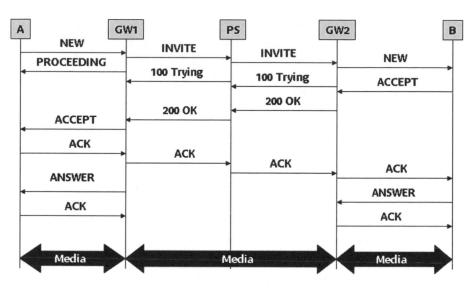

Figure 16.7 Call establishment

FQDN (Fully Qualified Domain Name) was provided to **A** prior to the call establishment process. Several methods may be used to provide such configuration information. For instance, a static configuration may be envisaged, or dynamic methods can be activated. Indeed, DHCP (Dynamic Host Configuration Protocol, [DHCP]), SLP (Service Location Protocol, [SLP]) or IAX may be used to provision customer devices with appropriate service-related configuration data. This configuration data mainly encloses the contact address (either an IP address or an FQDN) of the service contact point, an NTP (Network Time Protocol) or DNS (Domain Name Service) server and so on.

- The **NEW** message is then routed to **GW1**. Once received, the IAX–SIP interworking function is invoked and an **INVITE** message is generated. The generated **INVITE** message only uses the identities of 'Called' and 'Caller' URIs. For the remaining parts of the message, local information (that of **GW1**) is used. Since no error is experienced, this message is forwarded to the core service nodes; more precisely, to **PS**. In the meantime, a **PROCEED-ING** message is sent to **A** to notify it that the session request is currently being handled by the service.
- The **INVITE** message is routed to **PS**. Once it is received, **PS** executes its classical SIP logic. A lookup request is issued to the registration database in order to retrieve the contact point of the called party. In this example, an AoC of **GW2** is provided. The **INVITE** is consequently forwarded to **GW2**. A **100 Trying** message is also sent to **GW1**.
- Upon receipt of the **INVITE** message by **GW2**, the SIP–IAX interworking function is invoked. As a consequence, a **NEW** message is generated. That message is forwarded to the AoC of **B**. A **100 Trying** message is also sent to **PS**.
- **B** accepts the call request and issues an **ACCEPT** message. This is routed to **GW2**, which invokes its IAX–SIP interworking function and generates a **200 OK** message.
- The **200 OK** is relayed to **GW1** via **PS**. **GW1** invokes in its turn an **ACCEPT** message and sends it to **A**.

- **A** issues an **ACK** message to **GW1**. An SIP **ACK** message is then generated by **GW1** and routed to **GW2** via **PS**. In the meantime, an **ANSWER** message is generated by **GW1** and is sent to **A**. An **ACK** message is then generated by **A** and sent back to **GW1**.
- Once the SIP **ACK** message is received by **GW2**, an IAX **ACK** message is generated and sent to **B**. **B** replies with an **ANSWER** message. Once this is received by **GW2**, an IAX **ACK** message is sent back to **B**.
- At this stage, IAX media messages can be exchanged between **A** and **GW1**, and **B** and **GW2**. These messages are transcoded to RTP streams, which are exchanged between **GW1** and **GW2**.

This implementation alternative suffers from a synchronisation problem, as described in Section 16.3.5.3.

16.3.5.2 Basic Procedure with Authentication

When authentication is enforced to place a call, additional messages are exchanged so long as the service platform ensures that the called party is allowed to invoke the service, as shown in Figure 16.8.

In this context, the following steps are taken:

- A **NEW** message is issued by **A** to initiate a call to **B**. This message is sent to the service contact point **GW1**.
- The **NEW** message is then routed until it is delivered to **GW1**. Once it is received by **GW1**, the IAX–SIP interworking function is invoked. Then an **INVITE** message is generated. This

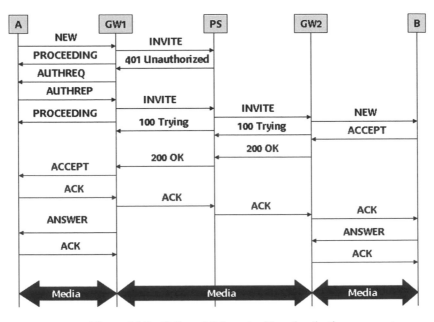

Figure 16.8 Call establishment with authentication

generated **INVITE** message only uses the identities of 'Called' and 'Caller' URIs. For the remaining parts of the message, local information (that of **GW1**) is used. Since no error is experienced, this message is forwarded to the core service nodes; more precisely, to **PS**. In the meantime, a PROCEEDING message is sent to **A** to notify it that the session request is currently being handled by the service.

- The **INVITE** message is routed to **PS**. Once received, **PS** executes its classical SIP logic. Since authentication is required, a **401 Unauthorized** is issued by **PS** and sent back to **GW1**.
- Once it is received by **GW1**, the interworking function is invoked and an **AUTHREQ** is generated and sent to **A**. The **AUTHREQ** mainly encloses a security challenge.
- Upon receipt of this message by **A**, a security hash is computed based on the security challenge included in the **AUTHREQ**. This security hash is then conveyed to **GW1** in an **AUTHREP** message.
- Once it is received by **GW1**, the IAX–SIP Interworking function is invoked. Consequently, an INVITE message is generated. This generated **INVITE** message only uses the identities of 'Called' and 'Caller' URIs and includes the security hash. In the meantime, a **PROCEED-ING** message is sent to **A** to notify it that the session request is currently being handled by the service.
- **The PROCEEDING** message is received by **PS**, which performs a lookup request based on its registration database to retrieve the contact point of the called party. In this example, an AoC of **GW2** is provided. The **INVITE** is consequently forwarded to **GW2**. A **100 Trying** message is also sent to **GW1**.
- Upon receipt of the **INVITE** message by **GW2**, the SIP–IAX interworking function is invoked. As a consequence, a **NEW** message is generated. That message is forwarded to the AoC of **B**. A **100 Trying** message is also sent to **PS**.
- **B** accepts the call request and issues an **ACCEPT** message. This is routed to **GW2**, which invokes its IAX–SIP interworking function and generates a **200 OK** message.
- The **200 OK** is relayed to **GW1** via **PS**. **GW1** invokes in its turn an **ACCEPT** message and sends it to **A**.
- **A** issues an **ACK** message to **GW1**. An SIP **ACK** message is then generated by **GW1** and routed to **GW2** via **PS**. In the meantime, an **ANSWER** message is generated by **GW1** and is sent to **A**. An **ACK** message is then generated by **A** and sent back to **GW1**.
- Once this is received by **GW2**, an IAX **ACK** message is generated and sent to **B**. **B** replies with an **ANSWER** message. Once this is received by **GW2** an IAX **ACK** message is sent back to **B**.
- At this stage, IAX media messages can be exchanged between **A** and **GW1**, and **B** and **GW2**. These messages are transcoded to RTP streams that are exchanged between **GW1** and **GW2**.

16.3.5.3 Synchronisation Issue

The implementation solution described in Sections 16.3.5.1 and 16.3.5.2 suffers from a synchronisation issue caused by the difference between IAX and SIP in terms of session establishment. Since IAX messages are sent over UDP, all control messages are reliable. This means that an **ACK** message is issued to acknowledge the receipt of an IAX message. Furthermore, IAX session establishment can be decomposed into two phases: the **NEW-ACCEPT** phase and the **ANSWER-ACK** one. The first phase is similar to the SIP procedure to

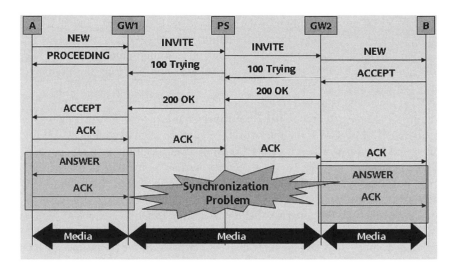

Figure 16.9   Call establishment: synchronisation issue

set up calls. As a consequence of these distinct ways of setting up sessions, a synchronization issue may arise as a result of IAX call legs not being successfully established. Figure 16.9 shows when this may occur.

Figure 16.10 shows an example of a failed call- establishment session. In this example, two call legs have been successfully established: **B–GW2** and the SIP one between **GW1** and **GW2**. Unidirectional RTP streams are exchanged between **GW1** and **GW2** even if the IAX call leg between **A** and **GW1** has failed. **GW1** retransmits the **ANSWER** message several times because no **ACK** message has been received from **A**.

To avoid this synchronisation issue, we propose to introduce a new procedure which aims at ensuring that all call legs are successfully established. IAX–SIP interconnection nodes host an additional function, which is defined as follows:

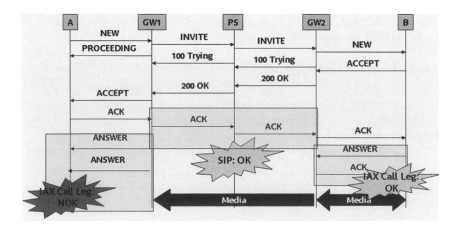

Figure 16.10   Synchronisation issue: example of a call flow

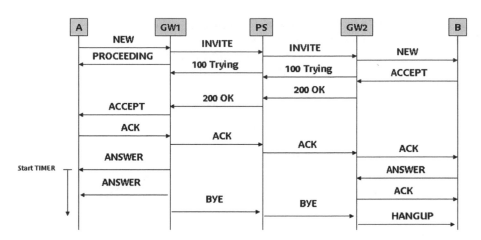

Figure 16.11   Call flow example to illustrate a synchronisation solution

- Introduce a new timer denoted **IAX_SIP_TIMEOUT**. The value of this timer is configured by each service provider on the IAX–SIP interconnection node.
- If no **ANSWER** or **ACK** message is received from the user agent before **IAX_SIP_TIME-OUT** expires, the ongoing session is considered as failed.
- The IAX–SIP interconnection node must issue a SIP **BYE** message to the service platform to notify it that this call leg has failed.

Figure 16.11 reproduces the messages that are exchanged when the aforementioned solution is enforced.

16.3.6 Call Tear-Down

Figure 16.12 shows the exhange of messages that must occur between the involved parties in order to tear down an ongoing call. Concretley,

- **A** sends a **HANGUP** message to its service contact point **GW1**.

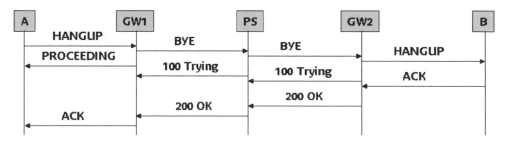

Figure 16.12   Terminating an ongoing call

- Once it is received by **GW1**, the IAX–SIP interworking function is invoked. As a result, a **BYE** message is built and then sent to **PS**.
- This **BYE** message is relayed by **PS** and delivered to **GW2**.
- Upon receipt of this message by **GW2**, the IAX–SIP interworking function is invoked. A **HANGUP** message is generated and then sent to **B**.
- Once it is received by **B**, an **ACK** message is routed back to **A**.

16.3.7 Aliveness of Registered Users

Unlike in pure SIP deployment scenarios, the use of IAX at the access segment allows the service platform to ensure that a remote user agent is still alive. **POKE** messages are sent regularly.

16.3.8 Registration Release without Authentication

Figure 16.3 illustrates the exchange of messages that occurs when a user releases its registration to a service without the support of authentication.

As shown in Figure 16.13, the following steps are taken to allow a given user a gent **A** to unsubscribe from a service:

- Because **A** is provisioned with a service contact point for service registration purposes, **A** sends its unsubscription message by invoking an IAX **REGREL** message. This message is routed to **GW1**.
- Once it is received by **GW1**, the interworking function is invoked. An SIP **REGISTER** message is built. This message encloses a **CONTACT** header with an **EXPIRES** field positioned to **0**. In SIP practices, this means that the remote user wants to deregister from the service. The **REGISTER** message is then sent to **R**.
- Upon receipt of the **REGISTER** message by **R**, appropriate functions and operations are invoked in order to deregister **A** from the service.
- To indicate to the remote user that its request has been successfully handled, a **200 OK** message is sent by **R** to **GW1**.
- This message is received by **GW1** and the SIP–IAX interworking function is invoked. As a result, an IAX **REGACK** is built. This message is then sent to **A**.
- To terminate the session, **A** sends an IAX **ACK** message.
- Once it is received by **GW1**, all states related to **A** as maintained by **GW1** are released.

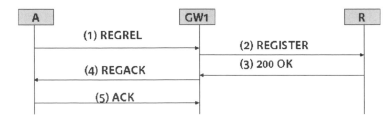

Figure 16.13   Call flow example of registration release without authentication

This example shows that a user can unsubscribe from a service in a transparent manner. The complexity related to the coexistence of IAX and SIP in the delivery of the service is hidden from end users.

16.3.9 Registration Release with Authentication

The example illustrated in Figure 16.13 is valid only if no authentication procedures are activated. From an operational standpoint, service requests should be authenticated and identity-checked. Indeed, service providers should put in place adequate means to verify that a given user is authorised to issue a particular message. Furthermore, the content of issued messages should be controlled. One user must not be able to unsubscribe another from the service. Figure 16.14 illustrates the exchange of messages that occurs when a user **A** releases its registration to a service with authentication procedures activated.

As shown in Figure 16.14, the following steps must be taken in order for **A** to unsubscribe from the service:

- **A** sends its unsubscription message by invoking an IAX **REGREL** message. This message is routed to **GW1**.
- Once it is received by **GW1**, the interworking function is invoked. An SIP **REGISTER** message is built. This message encloses a **CONTACT** header with an **EXPIRES** field positioned to **0**. In SIP practices, this means that the remote user wants to deregister from the service. The **REGISTER** message is then sent to **R**.
- Upon receipt of the **REGISTER** message by **R**, appropriate functions and operations are invoked to verify whether a remote user agent is authorised to issue this request. A **401 Unauthorized** message is sent to **GW1**.
- Once it is received by **GW1**, the SIP–IAX interworking function is invoked. Then a **REGAUTH** message is built. This message is sent to **A**. The **REGAUTH** must include a security challenge.
- Once it is received by **A**, a **REGREL** is resent by **A** to **GW1**. The **REGREL** must enclose a security hash.
- This second **REGREL** is then sent to **GW1**. As with the previous **REGREL** message, the IAX–SIP interworking function is invoked. A **REGISTER** message with a security hash is built. This message is forwarded to **R**.

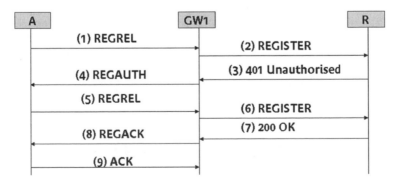

Figure 16.14 Call flow example of a registration release with authentication

• Upon receipt of this message, **R** verifies that the security negotiation has succeeded and proceeds to unsubscribe from the service. To indicate to the remote user that its request has been successfully handled, a **200 OK** message is sent by **R** to **GW1**.
• This message is received by **GW1** and the SIP–IAX interworking function is invoked. As a result, an IAX **REGACK** is built. This message is then sent to **A**.
• To terminate the session, **A** sends an IAX **ACK** message.
• Once it is received by **GW1**, all states related to **A** as maintained by **GW1** are released.

At the end of this process, **A** is unsubscribed from the service.

16.4 Bandwidth Optimisation: An Extension to SIP

Section 16.3.5 describes the procedure that must be followed to place telephony calls (or other session-based services) between remote user agents. Once signalling messages are exchanged, media streams are relayed by services nodes, especially the IAX–SIP interconnection node, as illustrated in Figure 16.15.
  In this implementation scenario:

• IAX is used to convey media streams in IAX call legs.
• RTP is used to send/receive media flows between IAX–SIP interconnection nodes.

This implementation option suffers from two drawbacks:

• Bandwidth is not optimised at the service domain, since RTP requires more bandwidth than the IAX media coding scheme.
• Since direct communication is enforced between access nodes, the activation of the IAX–RTP interworking function consumes resources and may induce additional delays when delivering media streams to end users.

These drawbacks can be avoided if IAX is also used to send media streams between access nodes. In this context, the media of all call legs will be conveyed using IAX as the sole media protocol, as illustrated in Figure 16.16.
  In order to implement this optimised procedure, SIP is used to initiate an IAX media session. Concretely, SDP (Session Description Protocol, [SDP]) should be extended to include IAX as the media protocol. SDP media lines which describe CODECs are also used for IAX

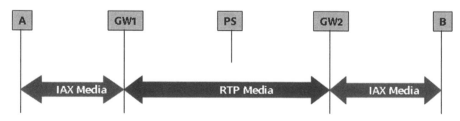

Figure 16.15   Media flows

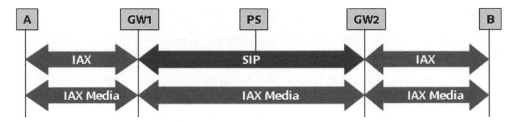

Figure 16.16   IAX: end-to-end media protocol

(Figure 16.17). Furthermore, SDP offers must include a **Source Call Number** for each call participant. Thus, an SDP offer may be organised as follows:

- Protocol: IAX
- Connection IP Address: local IP address
- Port Number: IAX port number
- List of Supported CODECs
- Source Call Number.

16.5 Conclusion

This chapter provided several call flows to illustrate the behaviour of the proposed IAX–SIP interworking function. The introduction of such a function into operational networks should mainly ease the traversal of middleboxes. Several technical points should be investigated further, mainly the deployment of advanced services. The IAX–SIP interworking function should be standardised and open so as to allow a rapid specification effort and therefore increase the availability of commercial products.

This chapter also introduced an extension to SDP to allow end-to-end bandwidth optimisation. The deployment of such a feature can be incremental.

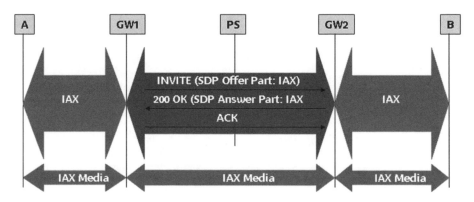

Figure 16.17   IAX extension to SDP

References

[DHCP] Droms, R., 'Dynamic Host Configuration Protocol', RFC 2131, March 1997.

[RTP] Schulzrinne, H., Casner, S., Frederick, R., and Jacobson, V., 'RTP: A Transport Protocol for Real-Time Applications', RFC 3550, July 2003.

[SDP] Handley, M., Jacobson, V., and Perkins, C., 'SDP: Session Description Protocol', RFC 4566, July 2006.

[SIP] Rosenberg, J., Schulzrinne, H., Camarillo, G., Johnston, A., Peterson, J., Sparks, R. et al., 'SIP: Session Initiation Protocol', RFC 3261, June 2002.

[SLP] Guttman, E., Perkins, C., Veizades, J. and Day, M., 'Service Location Protocol, Version 2', RFC 2608, June 1999.

17

Validation Scenario

17.1 Overview

The main purpose of this chapter is to provide some details about a deployment scenario assessing the validity of the approach proposed in Chapter 16.

One way of assessing the validity of the conclusions in this book is to conduct lab-based experiments. To do so, we propose to use Asterisk PBX (*) as the IAX-SIP gateway and SER as the SIP proxy server. IAX channels can be configured at the access side and SIP channels between the Asterisk server and the SER ones. Figure 17.1 represents this configuration.

The SER module is available at www.iptel.org/ser and Asterisk at www.asterisk.org. SER should be declared as peer in both Asterisk servers. IAX users must be configured in the attached Asterisk server. Additional information related to the required configuration is provided in the following sections.

17.2 Configuring Asterisk Servers

17.2.1 Configuration Operations

Asterisk uses the concept of 'channel' to denote a 'connection which brings in a call to the Asterisk PBX'.[1] For instance, an Asterisk channel could be a connection to an ordinary telephone handset or to a soft phone. Examples of supported Asterisk channels include:

- H.323
- IAX and IAX2
- MGCP (Media Gateway Control Protocol)
- SIP
- etc.

Note that Asterisk does not make any distinction between 'telephone lines' and 'telephones'.

---

[1] www.ip6net.net/voip-info.org/wiki/view/Asterisk + channels.html.

Inter-Asterisk Exchange (IAX): Deployment Scenarios in SIP-Enabled Networks   Mohamed Boucadair
© 2009 John Wiley & Sons, Ltd

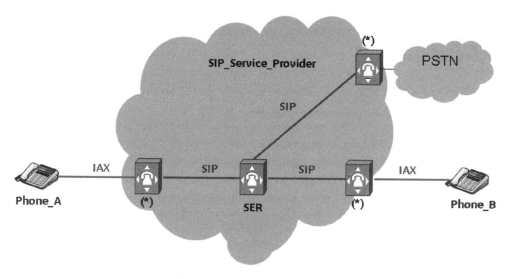

Figure 17.1   Validation scenario

Within this chapter, two channels are required to be configured to set up the validation scenario illustrated in Figure 17.2. Indeed, two channels must be configured in both Asterisk servers: the IAX and SIP channels.

- The SIP channel is used to manage connections with the SIP proxy server.
- The IAX channel is used to manage connections with end-users' phones.

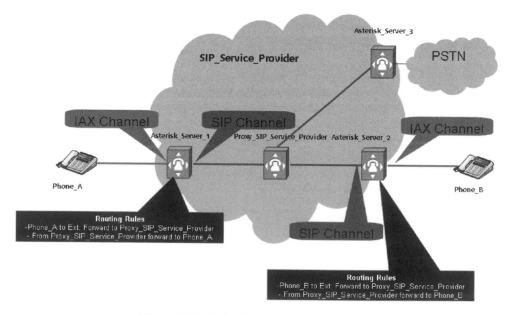

Figure 17.2   IAX-SIP configuration operations

In addition to the configuration of these two channels, both Asterisk servers should be configured according to the following policies:

- For incoming call requests (received in the context of the SIP channel; that is, from **Proxy_SIP_Service_Provider**), route the call to the IAX device.
- For received call requests from the IAX phone, forward the request to **Proxy_SIP_Service_Provider**.

In order to enforce these policies and to configure the aforementioned channels, three configuration files must be used:

- **sip.conf**: used to provide parameters and configuration data related to SIP-based communications.
- **iax.conf**: used to provide parameters and configuration data related to IAX-based communications.
- **extension.conf**: used to drive the routing process of received call requests.

Below, excerpts from these files are provided for illustration purposes.

17.2.2 Configuration Files

17.2.2.1 sip.conf

Asterisk_Server_1
Asterisk_Server_1 is configured to send **REGISTER** requests to the SIP-based service. A user ID and a password are provided in the configuration file. Both incoming and outgoing calls are configured as shown in Table 17.1.

Asterisk_Server_2
The configuration of **Asterisk_Server_2** is similar to that of **Asterisk_Server_2**. Table 17.2 provides an excerpt from the **sip.conf** file. The configuration lines shown indicate to the server that the SIP service should be registered using the user ID and password indicated in the file.

17.2.2.2 iax.conf

Asterisk_Server_1
In addition to SIP-based configuration, the IAX channel must be configured. Table 17.3 provides an excerpt from the **iax.conf** file. This file indicates that an IAX-based user agent is allowed access to the Asterisk server. The **host** parameter is positioned to 'dynamic' because **Phone_A** may not have a static IP address.

Asterisk_Server_2
A similar configuration is enforced in **Asterisk_Server_2**, as indicated in Table 17.4.

Table 17.1   Excerpt from **sip.conf** file of **Asterisk_Server_1**

```
[ general]
;
; Other configuration parameters
;
register => user_id_A:passwdservicesipidA@service-sip.com
[outgoing_calls]; Configure outgoing calls
type=peer
host= proxy-service-sip.com
user= user_id_A
secret= passwdservicesipidA
fromuser= user_id_A
fromdomain= service-sip.com
nat=no
canreinvite=no
[incoming_calls]; Configure incoming calls from the SIP Service Provider
type=peer
host= service-sip.com
context=incoming_calls
nat=no
canreinvite=no
;
; Other configuration parameters
;
```

17.2.2.3 extension.conf

In order to enforce routing policies at Asterisk servers when receiving a call request, **extension. conf** should be configured appropriately. Tables 17.5 and 17.6 provide excerpts from the files as configured on both Asterisk servers.

Asterisk_Server_1
Asterisk_Server_1 should route all received calls from **Phone_A** to the SIP service platform. Moreover, all received calls from the SIP service must be forwarded to **Phone_A**, as indicated in Table 17.5.

Asterisk_Server_2
Like **Asterisk_Server_1**, **Asterisk_Server_2** is configured to route received calls from the SIP service platform to **Phone_B**, and all those issued by **Phone_B** to the SIP service.

17.3 Configuring the SIP Express Router (SER)

17.3.1 Overview

Figure 17.3 highlights the routing policies which require to be configured on the SIP proxy server. SER must be configured so as to route received calls to their destinations. Concretely, calls to **Phone_A** should be forwarded to **Asterisk_Server_1**, those destined to **Phone_B**

Table 17.2   Excerpt from **sip.conf** file of **Asterisk_Server_2**

```
[general]
;
; Other configuration lines
;
register => user_id_B:passwdservicesipidB@service-sip.com
[outgoing_calls]; Configure outgoing calls
type=peer
host= proxy-service-sip.com
user= user_id_B
secret= passwdservicesipidB
fromuser= user_id_B
fromdomain= service-sip.com
nat=no
canreinvite=no
[incoming_calls]; Configure incoming calls from the SIP Service Provider
type=peer
host= service-sip.com
context=incoming_calls
nat=no
canreinvite=no
;
; Other configuration parameters
;
```

should be routed to **Asterisk_Server_2**, and those to other PSTN numbers should be sent to **Asterisk_Server_3**.

In order to enforce these policies, **ser.cfg** of **Proxy_SIP_Service_Provider** should be appropriately configured.

17.3.2 Configuration File

Table 17.7 provides an excerpt of the **ser.cfg** of the **Proxy_SIP_Service_Provider**.

Table 17.3   Excerpt from **iax.conf** file of **Asterisk_Server_1**

```
;
; Other configuration parameters
;
[Phone_A]; Configure an IAX account for Phone_A
type=friend
secret=local_passwd
host=dynamic
context=local
;
; Other configuration parameters
;
```

Table 17.4   Excerpt from **iax.conf** file of **Asterisk_Server_2**

```
;
; Other configuration parameters
;
[Phone_B]; Configure an IAX account for Phone_B
type=friend
secret=local_passwd
host=dynamic
context=local
nat=yes
;
; Other configuration parameters
;
```

Table 17.5   Excerpt from **extension.conf** file of **Asterisk_Server_1**

```
;
; Other configuration commands
;
[incoming_calls]; received SIP calls are forwarded to phone_A
exten => s,1,Dial(IAX/phone_A)
[local]; All outgoing calls are forwarded to SIP Service Provider
exten => _X.,1,Dial(SIP/outgoing_calls/${EXTEN})
```

Table 17.6   Excerpt from **extension.conf** file of **Asterisk_Server_2**

```
;
; Other configuration commands
;
[incoming_calls]; received SIP calls are forwarded to phone_B
exten => s,1,Dial(IAX/phone_B)
[local]; All outgoing calls are forwarded to SIP Service Provider
exten => _X.,1,Dial(SIP/outgoing_calls/${EXTEN})
```

17.4 User Agent Configuration

As a final step, end-user devices are configured to send all their service requests to a service contact point. Indeed, and as illustrated in Figure 17.4, **Phone_A** (respectively **Phone_B**) is configured to send registration and signalling messages to **Asterisk_Server_1** (respectively **Asterisk_Server_2**).

17.5 Conclusion

This chapter has shown a validation scenario to assess the feasibility of the proposed strategy for the introduction of IAX into an SIP-enabled environment. This validation scenario does not

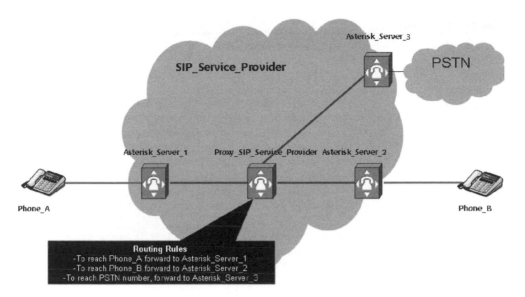

Figure 17.3   SER configuration operations

Table 17.7   Excerpt from the **ser.cfg** file of the **SIP_Proxy_Service_Provider**.

```
#
# Other configuration Parameters
#
# Route received call request
route[INBOUND]
{
    # lets see if know the callee
    if (lookup_user("$tu.uid", "@ruri")) {
        if ($tu.fwd_always_target) {
            altr2uri("$tu.fwd_always_target");
            route(FORWARD);
        }
        # native SIP destinations are handled using our USRLOC DB
        if (lookup_contacts("location")) {
            append_hf("P-hint: usrloc applied\r\n");
            # we set the TM module timers according to the prefences
            # of the callee (avoid too long ringing of his phones)
            # from the FAILURE_ROUTE below
            if ($t.fr_inv_timer) {
                if ($t.fr_timer) {
                    t_set_fr("$t.fr_inv_timer", "$t.fr_timer");
                } else {
                    t_set_fr("$t.fr_inv_timer");
                }
            }
            route(FORWARD);
        } else {
            sl_reply("480", "User temporarily not available");
            drop;
```

(continued)

Table 17.7 (*Continued*)

```
            }
        }
}
# Route received call request to Asterisk_Server_3
route[PSTN]
{
    # Only if the AVP 'gw_ip' is set and the request URI contains
    # only a number we consider sending this to the PSTN GW.
    # Only users from a local domain are permitted to make calls.
    # Additionally you might want to check the acl AVP to verify
    # that the user is allowed to make such expensives calls.
    if ($f.did && $gw_ip &&
        uri=~"sips?:\+?[0-9]{3,18}@.*") {
        # probably you need to convert the number in the request
        # URI according to the requirements of your gateway here
        # if an AVP 'asserted_id' is set we insert an RPID header
        if ($asserted_id) {
            xlset_attr("$rpidheader",
"<sip:%$asserted_id@%@ruri.host>;screen=yes");
            replace_attr_hf("Remote-Party-ID", "$rpidheader");
        }
        # just replace the domain part of the RURI with the
        # value from the AVP and send it out
        attr2uri("$gw_ip", "domain");
        route(FORWARD);
    }
}
```

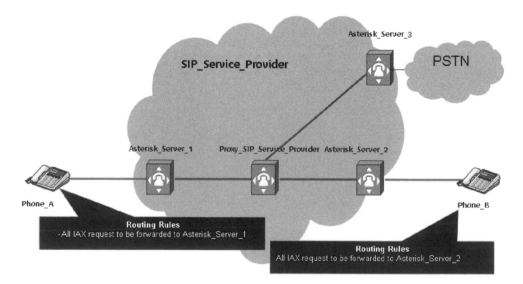

Figure 17.4   User agent configuration

aim to assess the performance of the proposed solution but only to provide a 'proof of concept' system.

Further investigation should be undertaken and IAX–SIP interworking function should be specified in depth and implemented. System-oriented testing should then be carried out, and analysis achieved.

Further Reading

Asterisk platform, http://www.asterisk.org/.
SIP Express Router, http://www.iptel.org/ser/.

Index

3GPP: 1

Access Control: 118, 143, 202, 203, 207, 211, 214

Access segment: 4, 7, 127, 129, 132, 193, 195, 197, 204, 210, 218, 220, 223, 225, 226, 234

Application-Level Gateway: 3

Asterisk: 239, 240, 241, 242

Asterisk Servers: 239, 240, 241, 242

Authentication: 4, 6, 13, 79, 85, 92, 93, 94, 95, 96, 97, 107, 112, 122, 128, 169, 192, 209, 217, 226, 227, 228, 230, 231, 234, 235

Auto Answer: 56

Bandwidth Optimisation: 236

Border Equipment: 210

Call Flow: 10, 89, 91, 93, 95, 98, 99, 100, 101, 122, 177, 205, 223, 225, 226, 237

Call Monitoring: 8, 17, 89, 99, 101

Call Number: 54

Call Optimisation: 89, 100, 101, 130, 180, 185, 187

Call Setup: 8, 12, 15, 17, 43, 89, 96, 97, 98, 101, 103, 107, 117, 122, 141, 142, 166, 180, 182, 187, 188, 226, 228

Call tear-down: 89, 99, 123, 226, 233

Call Transfer: 100, 101, 129

Called Context: 49

Called Number: 48

Calling Name: 49

Calling Number: 48

Calling TNS: 60

Calling TON: 60

Capability Mismatches: 200, 203, 211

CAPEX: 4, 148, 155, 175, 191

Cause Code: 61

CODEC: 8, 13, 39, 41, 42, 50, 62, 66, 69, 70, 72, 73, 77, 103, 105, 106, 107, 108, 109, 110, 176, 200, 216, 236, 237

Connectivity: 2, 4, 8, 12, 17, 20, 22, 79, 80, 84, 87, 93, 100, 101, 175, 176, 177, 184, 185, 186, 207, 212

Contact Table: 92, 93, 94, 120, 122, 160, 161, 162, 165

Decoupled Protocol: 14, 126, 127, 211

Demultiplexing: 12, 80, 81

Device Type: 58

DP Status: 54

DUNDi: 15, 118, 119, 122, 125, 126, 131, 209

Encrypted Frames: 35, 44

Encryption Key: 60

Encryption: 6, 14, 44, 46, 85, 112, 118, 202, 203, 209, 211, 214, 217, 220

End-to-End: 2, 3, 8, 10, 15, 27, 103, 117, 118, 122, 127, 133, 169, 175, 198, 203, 205, 217, 237

ENUM: 8, 15, 17, 19, 24, 25, 26, 27, 28, 29, 31, 32

Firmware: 5, 6, 14, 58, 59, 69, 72, 73, 77, 89, 90, 91, 92, 101, 209, 226

Firmware download: 58, 69, 77, 90, 91

Firmware Update: 89, 91, 92, 101, 226

Fixed-Mobile Convergence: 213